e of
stic
usic
ntre

The Language of Electroacoustic Music

Edited by
Simon Emmerson

MACMILLAN
PRESS
Music Division

First published 1986

Published by
THE MACMILLAN PRESS LTD
Houndmills, Basingstoke, Hampshire RG21 2XS
and London
Companies and representatives
throughout the world

Typeset by Rowland Phototypesetting Ltd
Bury St Edmunds, Suffolk
in 10/13pt Sabon

Transferred to digital print on demand, 2003

Printed and bound by Antony Rowe Ltd, Eastbourne

British Library Cataloguing in Publication Data

The Language of electroacoustic music.
1. Electronic music – History and criticism 2. Concrete music – History and criticism
I. Emmerson, Simon
789.9 ML1092

ISBN 0-333-39759-2
ISBN 0-333-39760-6 Pbk

Contributors

Pierre Boulez is Director of the Institut de Recherche et Coordination Acoustique/Musique (IRCAM) in Paris.

Simon Emmerson is director of the Electroacoustic Music Studio at The City University, London.

Jonathan Harvey is Professor of Music at Sussex University.

David Keane is Professor of Music at Queen's University, Kingston, Ontario.

Tod Machover is Assistant Professor of The Media Laboratory, Massachusetts Institute of Technology.

Michael McNabb is a composer working at the Center for Computer Research in Music and Acoustics at Stanford University.

Bruce Pennycook is Assistant Professor of music at Queen's University, Kingston, Ontario.

Denis Smalley is director of the Electroacoustic Music Studio at University of East Anglia.

Barry Truax is Associate Professor in the Department of Communication, Simon Fraser University, British Columbia.

Trevor Wishart is a freelance composer based in York, UK.

Contents

Introduction

Simon Emmerson

Few genres in Western music have leapt fully fledged from the heads of their creators. It has usually been possible to identify antecedents and influences in each case: a reflection of our desire to contextualize the revolutionary and thus reassure ourselves of the continuity of our traditions. But in no age has this process of analysis and criticism been so immediate; at no time has the new been so immediately the subject of scrutiny in a range of publications from the popular press to the esoteric research paper.

Immediate academicization is dangerous, especially when applied to artistic forms; all information is afforded a veneer of neutrality, all events treated as equally worthy of analysis; webs of historic reference are established: who was first or who influenced whom; even though details of a style or technique may be examined, no real evaluation of the music is made and the emergence of a true critique is stifled. Much of this nature has been written on electroacoustic music, the only truly original development of Western music in the 20th century. Such writing has covered two areas: analysis of the musical techniques and analysis of the technical means. Little has been written – at least in English – on the musical aims, the ethic and aesthetic of the music. As an example we may cite the currently accepted 'history' of the 1950s. The simplistic view that the great debate between *musique concrète* and *elektronische Musik* was about materials alone misses much about the motivations. This point is not merely academic, it influences how we teach and compose today.

This book seeks to contribute to a slower process than that of merely presenting the latest facts on music or technology; it seeks to initiate a debate aimed at establishing a longer-term view, a clearer picture of where we stand at the meeting point of two millennia. While it encompasses a variety of approaches and arguments, I would stress several which the contributors have in common. The first of these is a commitment to an experimental tradition, one which seeks to find new solutions, to test, to research and to modify assumptions accordingly. This distinguishes the 1980s from the 1960s during which the liberation of sound was hardly the subject of rigorous assessment. The clue to the difference between the two decades lies in the recognition of the concept of failure: failure contributes to discovery, providing it is recognized and is allowed to motivate the modification of one's utterances, one's language. The second shared approach – and all the contributors are composers – is in the commitment to communication and an acceptance of the proper role of the listener's perception in the musical process. It is true that a wide range of relationships between 'pre-composition' and 'aural composition' are to be found, but the listener remains the arbiter. There are, of course, dissenters, but the debate has only just begun to establish an agreed vocabulary upon which discourse may be based.

It must not be thought that such views merely pander to an audience; electroacoustic music would have become absorbed completely into popular electronic music forms long ago had this been the case. Electroacoustic music challenges the listener in two fundamental ways. In the case of tape music it asks that the imagination replace the visual stimulus of live performance – something which radio initiated and which is paradoxically reinforced in popular music video. It also allows little recourse to another visual cue: the score. Scores for electroacoustic music exist, of course, but have a variety of functions: for performance with instruments or over complex loudspeaker systems, or simply for reading and background information, but rarely for the re-creation of the work itself. The score moves away from prescription towards description. The emphasis is in all cases away from a reliance upon the written hieroglyph as a means to express, or at least to transmit, musical utterance.

The book opens with the classic article by Pierre Boulez, 'Technology and the Composer', written in 1977 but still relevant today. In that it foreshadows many of the developments about which the other contributors write, it forms a Preface to the book. When Boulez wrote his article, computers had been used in music synthesis for little more than a decade. He felt the prevalent musical attitude to be one of 'conservative historicism' and pointed to the need to adopt a more progressive approach which incorporated technological development. Furthermore, he called for understanding and communication between composers and technologists, for: "a common language which would take account of the imperatives of musical invention and the priorities of technology". He felt both the technological resources and the musical concept to be fundamentally impoverished, and pointed to the need for a truly dialectical relationship between material and idea. It should be obvious from the contributions to this book that, with some reservations, such a reciprocal relationship is slowy developing. The fundamental question which Boulez considered it necessary to confront – "whether the material is adequate to the idea and the idea compatible to the material" – is now being asked.

The contributions to this volume have been grouped under three broad headings. Under the first – Materials and Language – my own chapter presents a context within which to discuss and categorize the musical material used by composers since the inception of electroacoustic music. Trevor Wishart seeks to establish commonly understood symbols within the sonic world which allow the direct transmission of ideas and the development of a common language within which unique and personal utterances can be made. Denis Smalley describes the 'spectro-morphological' approach to electroacoustic music, based entirely on the way sound-objects have, inherent within their perceptual qualities, the potential for building gestures, shapes and forms – indeed the potential for whole languages. His chapter is the first in English properly to examine precisely how these shapes are built up and may be developed.

The second section – Problems of Language – confronts some of the issues, aesthetic and practical, which have arisen from the rapid development of the technical resources. David Keane challenges the very use of the terms 'language' and indeed 'music' with respect to these developments, and concludes that new tools may alter the nature of the task in fundamental ways. Bruce Pennycook points to the failure of teaching institutions to continue to develop and use the technology which was in many cases initiated by them. He sees a divide between the potential for research in universities and the way in which the development of computer music systems is being affected by commercial pressures. This, in his opinion, is leaving electronic composers and students with possibly conflicting expressive musical languages.

In contrast, the last four chapters of the book bear out the fact that in certain well-funded centres creative work of a high degree of sophistication is being undertaken. In Boulez's terms, musical invention has appropriated the language of technology and has, in turn, begun to create its own musical language. The Influence of Computer Technology is rather a broad title for this final section in which four authors approach the topic in a very personal way. Michael McNabb's overview of the uses of the computer in music concludes with a description of a technique he is developing to shape and unify musical material. Barry Truax argues that too detailed an approach to the control of every musical parameter can result in a loss of gestural and expressive control and puts forward the case for computer music languages in which the user manipulates relatively few variables, each of which has a consequently greater effect. Jonathan Harvey, on the other hand, defends detailed parameter-by-parameter investigation and describes his own experience in timbre research and its stimulus to composition. Tod Machover discusses his own belief in a fundamental unifying principle in artistic creation and the ways in which computer technology is able to help him reflect this in his own music, and in particular in his forthcoming opera *Valis*.

The Language of Electroacoustic Music does not aim to be a comprehensive account of the many and varied manifestations of electroacoustic music today. It seeks to lay the foundations for discussion about aesthetic matters by clarifying the central issues. In this way it is hoped that the book will contribute to a genuine critique of the medium as it approaches the start of its fifth decade.

Simon Emmerson
London, February 1986

1
Technology and the Composer[1]

Pierre Boulez

Invention, in music, is often subject to prohibitions and taboos which it would be dangerous to transgress. Invention must remain the private, exclusive property of genius, or at least of talent. Indeed it is hard to find any purely rational explanation for it; by summoning up unpredictable results out of nothing it escapes analysis. But is this nothing really the total void appropriate to miracle-workers? And does the unpredictable come to exist in a totally unpredicted context? Invention cannot exist in the abstract, it originates in contact with music of the past, be it only the recent past; it exists through reflection on its direct or indirect antecedents. Such reflection concentrates naturally on the spiritual approach, the mental mechanisms and the intellectual development displayed by the work it takes as its models, but it concentrates also on the sound material itself, without whose support music cannot exist; musical material has evolved over the centuries, providing for each age a typical sound profile that is continually renewed – slowly perhaps, but inevitably.

Yet invention is faced today with a number of problems particularly concerned with the relation between the conception, we might even say the vision, of the composer and the realization in sound of his ideas. For some time now, the composer's mental approach, his 'wild' invention, has been free

[1] First published in English in the *Times Literary Supplement*, 6 May 1977

to follow very different paths from those that the medium, the sound material, can offer him. This divergence has caused blockages dangerous enough for invention to lose all its spontaneity; when either the material or the idea develops independently, unconcerned whether or not they coincide, a serious imbalance develops, to the detriment of the work, which is tugged this way and that, between false priorities. Underlying these blockages there are undoubtedly causes which are beyond the composer's power and over which he has little control, but of which he is – or should be – aware if he is to try to overcome them.

We think at once of blockages of a social kind. Since at least the beginning of this century, our culture has been oriented towards historicism and conservation. As though by a defensive reflex, the greater and more powerful our technological progress, the more timidly has our culture retracted to what it sees as the immutable and imperishable values of the past. And since a larger – though still limited – section of society has easier access to musical culture, having more leisure and spending power, and since modes of transmission have increased enormously and at the same time are cheaper, the consumption of music has considerably increased. This leads to a growing boredom with pieces that are frequently heard and repeated, and to a search for an alternative repertoire – one within the same radius of action as the well-known works and providing a series of substitutes for them. Only too rarely does it lead to a genuine broadening of the repertoire by giving fresh life to works which had become the exclusive property of libraries. The search for historical particularities of interpretation also serves to divert energies which are all too likely to be swallowed up by it. Thus the 'museum' has become the centre of musical life, together with the almost obsessive preoccupation with reproducing as faithfully as possible all the conditions of the past. This exclusive historicism is a revealing symptom of the dangers a culture runs when it confesses its own poverty so openly: it is engaged not in making models, nor in destroying them in order to create fresh ones, but in reconstructing them and venerating them like totems, as symbols of a golden age which has been totally abolished.

Among other consequences, an historicizing culture has almost completely blocked the evolution of musical instruments, which have come to a disastrous halt for both social and economic reasons. The great channels of musical consumption which exploit, almost exclusively, the works of the past consequently use the transmission appropriate to the past, when they were at their most effective. It is hardly necessary to add that this state of affairs is faithfully reflected in education, where the models selected for teaching are drawn from an extremely circumscribed period in the history of music, and consequently limit – from the outset – the techniques and sound material at the musician's disposal; even more disastrously, they give him a restricted outlook whereby his education becomes a definitive, absolute possession. The makers of musical instruments, having no vocation for economic suicide, meet the narrow demands made on them; they are interested only in fiddling about with established models and so lose all chance of inventing or transforming. Wherever there is an active market, in which economic demand has free play – in a field like pop music where there are no historical constraints – they become interested, like their colleagues who design cars or household appliances, in developing prototypes which they then transform, often in quite minimal ways, in order to find new markets or unexploited outlets. Compared with these highly prosperous economic circuits, those of so-called serious music are obviously impoverished, their hopes of profit are decidedly slender and any interest in improving them is very limited. Thus two factors combine to paralyse the material evolution of the contemporary musical world, causing it to stagnate within territory conquered and explored by other musical periods for their own and not necessarily our needs – the minimal extension of contemporary resources is thus restricted to details. Our civilization sees itself too smugly in the mirror of history; it is no longer creating the needs which would make renewal an economic necessity.

In another sector of musical life which has little or no communication with the 'historical' sect, the musical material itself has led a life of its own for the past thirty years or so, more or less independent from invention: out of revenge for its neglect and stagnation, it has formed itself into a surplus, and one wonders at times how it can be utilized. Its urgency expresses itself even before it is integrated into a theme, or into a true musical invention. The fact is that these technological researches have often been carried out by the scientifically minded, who are admittedly interested in music, but who stand outside the conventional circuit of musical education and culture. There is a very obvious conjunction here between the economic processes of a society which perpetually demands that the technology depending on it should evolve, and which devotes itself notoriously to the aims of storage and conservation, and the fall-out from that technology, which is capable of being used for sometimes surprising ends, very different and remote from the original research. The economic processes have been set to produce their maximum yield where the reproduction of existing music, accepted as part of our famous cultural heritage, is concerned; they have reduced the tendency to monopoly and the rigid supremacy of this heritage by a more and more refined and accessible technology.

Techniques of recording, backing, transmission, reproduction – microphones, loudspeakers, amplifying equipment, magnetic tape – have been developed to the point where they have betrayed their primary objective, which was faithful reproduction. More and more the so-called techniques of reproduction are acquiring an irrepressible tendency to become autonomous and to impress their own image of existing music, and less and less concerned to reproduce as faithfully as possible the conditions of direct audition; it is easy to justify the refusal to be faithful to an unrecorded reality by arguing that *trompe-l'oeil* reproduction, as it were, has little meaning given that the conditions of listening and its objectives are of a different order, that consequently they demand different criteria of perception. This, transposed into musical terms, is the familiar controversy about books and films on art: why give a false notion of a painting in relation to the original by paying exaggerated attention to detail, by controlling the lighting in an unusual way, or by introducing movement into a static world? . . . Whatever we make of this powerful tendency towards technological autonomy in the world of sound production, and whatever its motives or its justifications, one sees how rapidly the resources involved are changing, subject as they are to an inexorable law of movement and evolution under the ceaseless pressure of the market.

Aware of these forms of progress and investigation, and faced at the same time by stagnation in the world of musical instruments, the adventurous musical spirits have thought of turning the situation to their own advantage. Through an intuition that is both sure and unsure – sure of its direction, but unsure of its outcome – they have assumed that modern technology might be used in the search for a new instrumentation. The direction and significance of this exploration did not emerge until long after the need for it arose: irrational necessity preceded aesthetic reflection, the latter even being thought superfluous and likely to hamper any free development. The methods adopted were the outcome either of a genuine change of function, or of an adaptation, or of a distortion of function. Oscillators, amplifiers, and computers were not invented in order to create music; however, and particularly in the case of the computer, their functions are so easily generalized, so eminently transformable, that there has been a wish to devise different objectives from the direct one: accidental conjunction will create a mutation.

The new sound material has come upon unsuspected possibilities, by no means purely by chance but at least by guided extrapolation, and has tended to proliferate on its own; so rich in possibilities is it that sometimes mental categories have yet to be created in order to use them. To musicians accustomed to a precise demarcation, to a controlled hierarchy and to the codes of a convention consolidated over the centuries, the new material has proposed a mass of unclassified solutions, and offered us every kind of structure without any perspective, so affording us a glimpse of its immense potential without guidance as to which methods we should follow.

So we stand at the crossroads of two somewhat divergent paths: on the one hand a conservative historicism which, if it does not altogether block invention, clearly diminishes it by providing none of the new material it needs for expression, or indeed for regeneration. Instead, it creates bottlenecks, and impedes the circuit running from composer to interpreter, or, more generally, that from idea to material, from functioning productively; for all practical purposes, it divides the reciprocal action of these two poles of creation. On the other hand, we have a progressive technology whose force of expression and development are sidetracked into a proliferation of material means which may or may not be in accord with genuine musical thought – for this tends by nature to be independent, to the detriment of the overall cohesion of the sound world. (Having said which, one should note that long before contemporary technology, the history of musical instruments was littered with corpses: superfluous or over-complicated inventions, incapable of being integrated into the context demanded by the musical ideas of the age which produced them; because there was no balance between originality and necessity they fell into disuse.)

Thus inventors, engineers and technicians have gone in search of new processes according to their personal preferences, choosing this one or that purely by whim, and for fortuitous rather than for musically determined reasons – unless their reasons stemmed from their more exclusively scientific proccupations. But musicians, on the whole, have felt repelled by the technical and the scientific, their education and culture having in no way given them the agility or even the readiness to tackle problems of this kind. Their most immediate and summary reaction, therefore, is to choose from the samples available, or to make do at a level easily accessible to manipulation. Few have the courage or the means directly to confront the arid, arduous problems, often lacking any easy solution, posed by contemporary technology and its rapid development. Rather than ask themselves the double question, both functional and fundamental, whether the material is adequate to the idea and the idea compatible with the material, they give way to the dangerous temptation of a superficial, simple question: does the material satisfy my immediate needs? Such a hasty choice, detached from all but the most servile functions, certainly cannot lead far, for it excludes all genuine dialectic and assumes that invention can divorce itself from the material, that intellectual schemas can exist without the support of sound. This does not apply to the music of the past which was not, properly speaking, written for specified instruments, for its writing assumes absolutely the notion of the instrument, even of the monodic instrument within a fixed and limited register. If invention is uninterested in the essential function of the musical material, if it restricts itself to criteria of temporary interest, or fortuitous and fleeting coincidences, it cannot exist or progress organically; it utilizes immediate discoveries, uses them up, in the literal sense of the term, exhausting them without really having explored or exploited them. Invention thereby condemns itself to die like the seasons.

Collaboration between scientists and musicians – to stick to those two generic terms which naturally include a large number of more specialized categories – is, therefore, a necessity which, seen from outside, does not appear to be inevitable. An immediate reaction might be that musical invention can have no need of a corresponding technology; many representatives of the scientific world see nothing wrong with this and justify their apprehensions by the fact that artistic creation is specifically the domain of intuition, or the irrational. They doubt whether this utopian marriage of fire and water would be likely to produce anything valid. If mystery is involved, it should remain a mystery: any investigation, any search for a meeting-point is easily taken to be sacrilege. Uncertain just what it is that musicians are demanding from them, and what possible terrain there might be for joint efforts, many scientists opt out in advance, only seeing the absurdity of the situation: that is, a magus reduced to begging for help from a plumber! If, in addition, the magus imagines that the plumber's services are all that he needs, then confusion is total. It is easy to see how hard it will be ever to establish a common language for both technological and musical invention.

In the end, musical invention will have somehow to learn the language of technology, and even to appropriate it. The full arsenal of technology will elude the musician, admittedly; it exceeds, often by a big margin, his ability to specialize; yet he is in a position to assimilate its fundamental procedures, to see how it functions and according to which conceptual schemes – how far, in fact, it might or might not coincide with the workings of musical creation and how it could reinforce them. Invention should not be satisfied with a raw material come upon by chance, even if it can profit from such accidents and, in exceptional circumstances, enlarge on them. To return to the famous comparison, the umbrella and the sewing machine cannot create the event by themselves – it needs the dissecting-table too. In other words, musical invention must bring about the creation of the musical material it needs; by its efforts, it will provide the necessary impulse for technology to respond functionally to its desires and imagination. This process will need to be flexible enough to avoid the extreme rigidity and impoverishment of an excessive determinism and to encompass the accidental or unforeseen, which it must be ready later to integrate into a larger and richer conception. The long-term preparation of research and the instantaneous discovery must not be mutually exclusive, they must affirm the reciprocity of their respective spheres of action.

One can draw a parallel with the familiar world of musical instruments. When a composer learns orchestration, he is not asked to have either a practical, a technical or a scientific knowledge of all the instruments currently at our disposal. In other words, he is not expected to learn to play every one of these instruments, even if out of personal curiosity he may familiarize himself with one or other of them and even become a virtuoso. Furthermore, he is not expected to learn how the instruments were made, how they reached their present stage of development, by what means and along which path their history has evolved so that certain of their specific possibilities were stressed to the neglect of others and reflect on whichever aspect remains his personal choice. Still less is the composer expected to learn the acoustic structure of the sounds produced by a particular family of instruments; his curiosity or his general, extra-musical education may lead him to concern himself with these problems in so far as scientific analysis can confirm his impressions as a musician. He may have none of this literal knowledge, yet nothing in the functioning of an instrument, either practical, technical or scientific, should be beyond his understanding. His apprenticeship is in a sense not a real but a virtual one. He will know what is possible with an instrument, what it would be absurd to demand of it, what is simple and what is out of the question, its lightness or its heaviness, its ease of articulation or difficulty in sound production in various registers, the quality of the timbre, all the modifications that can be made either through technique itself or with the aid of such devices as the mute, the weight of each instrument, its relationship with the others; all these are things that he will verify in practice, his imagination abandoning itself to the delights of extrapolation. The gift lies in the grafting of intuition on to the data he has acquired. A virtual knowledge of the entire instrumental field will enable him to integrate into his musical invention, even before he actually composes, its vast hidden resources; that knowledge forms a part of his invention.

Thus a virtual understanding of contemporary technology ought to form part of the musician's invention; otherwise, scientists, technicians and musicians will rub shoulders and even help one another, but their activities will be only marginal one to the other. Our grand design today, therefore, is to prepare the way for their integration and, through an increasingly pertinent dialogue, to reach a common language which would take account of the imperatives of musical invention and the priorities of technology. This dialogue will be based as much on the sound material as on concepts.

Where the material is concerned, such a dialogue seems possible here and now: it offers an immediate interest and is far from presenting any insurmountable difficulties. From our education within a traditional culture we have learned and experienced how instrumental models function and what they are capable of. But in the field of electronics and computers – the instrument which would be directly involved – models do not exist, or only sporadically, and largely thanks to our imagination. Lacking sound schemes to follow, the new field seems exaggeratedly vast, chaotic, and if not inorganic at least unorganized. The quite natural temptation is to approach this new field with our tried and tested methods and to apply the grid of familiar categories which would seem to make the task easier and to which, for that reason, we would like to resort unthinkingly. The existing categories could, it is true, be helpful at first in mapping out virgin territory and enabling us, by reconstitution and synthesis, better to know the natural world which we think we know so well and which, the nearer we get to it, seems to elude the precision of our investigation. It is not only the question "what is a sound made of?" that we have to answer, but the much harder one of "how do we perceive this sound in relation to its constituent elements?" So by juxtaposing what is known with what is not known, and what is possible with what will be possible, we shall establish a geography of the sound universe, so establishing the continuity of continents where up until now many unknown territories have been discerned.

It goes without saying that the reasoned extension of the material will inspire new modes of thought; between thought and material a very complex game of mirrors is set up, by which images are relayed continuously from one to the other. A forceful, demanding idea tends to create its own material, and in the same way new material inevitably involves a recasting of the idea. We might compare this with architecture, where structural limitations have been radically changed by the use of new materials such as concrete, glass, and steel. Stylistic change did not happen overnight; there were frequent hesitations and references back to the past – to ennoble, as it were, these architectural upstarts. New possibilities triumphed over imitation and transformed architectural invention and concepts from top to bottom. These concepts had to rely much more than before on technology, with technical calculations intervening even in aesthetic choices, and engineers and architects were obliged to find a common language – which we are now about to set off to look for in the world of music.

If the choice of material proves to be the chief determinant in the development of creative ideas, this is not to say that ideas should be left to proceed on their own, nor that a change of material will automatically entail a revision of concepts relating to musical invention. Undoubtedly, as in the case of architecture, there will be caprices and hesitations, and an irrepressible desire to apply old concepts to the new material, in order to achieve – perhaps *ab absurdo*? – a kind of verification. But if we wish to pass beyond these immediate temptations, we shall have to strive to think in new categories, to change not only the methods but the very aim of creation. It is surprising that in the musical developments of the past sixty years many stylistic attitudes have been negative, their chief aim, need or necessity being to avoid referring back – if there has been such reference it has been produced in a raw unassimilated state, like a collage or parody, or even a mockery. In trying to destroy or amalgamate, reference in fact betrays the inability to absorb, it betrays the weakness of a stylistic conception unable to "phagocytose" what it takes hold of. But if one insists o stylistic integrity as a prime criterion, and if the material, through previous use, is rich in connotations, if it stimulates involuntary associations and risks diverting expression into unwanted directions, one is led in practice into playing, if not absolutely against the material, then at least to the limit of its possibilities. Coincidence no longer exists, or can only exist in the choice of a specialized area – in the rejection, that is, of many other areas which would impose references that were eccentric and too powerful. It would seem that this excessively cautious attitude could not persist in the face of new material from which connotations have been excluded; the relationship between idea and material becomes eminently positive and stylistic integrity is no longer at risk.

Creative thought, consequently, is in a position to examine its own way of working, its own mechanisms. Whether in the evolution of formal structures, in the utilization of determinism, or in the manipulation of chance, and whether the plan of assembly be based on cohesion or fragmentariness, the field is vast and open to invention. At its limits, one can imagine possible works where material and idea are brought to coincide by the final, instantaneous operation that gives them a true, provisional existence – that operation being the activity of the composer, of an interpreter, or of the audience itself. Certainly, the finite categories within which we are still accustomed to evolve will offer less interest when this dizzying prospect opens up: of a stored-up potential creating instant originality.

Before we reach that point, the effort will either be collective or it will not be at all. No individual, however gifted, could produce a solution to all the problems posed by the present evolution of musical expression.

Research/invention, individual/collective, the multiple resources of this double dialectic are capable of engendering infinite possibilities. That invention is marked more particularly by the imprint of an individual, goes without saying; we must still prevent this involving us in humdrum, particular solutions which somehow remain the composer's personal property. What is absolutely necessary is that we should move towards global, generalizable solutions. In material as in method, a constant flow must be established between modes of thought and types of action, a continual exchange between giving and receiving. Future experiments, in all probability, will be set up in accordance with this permanent dialogue. Will there be many of us to undertake it?

Materials and Language

2
The Relation of Language to Materials

Simon Emmerson

Mimetic and aural musical discourse

It is not the purpose of this chapter to delve yet again into the meaning of the term 'expression' used with respect to music. We will be concerned with one aspect of music perception brought to the fore once more by the development of electroacoustic music on tape: namely the possible relation of the sounds to associated or evoked images in the mind of the listener. The term 'image' may be interpreted as lying somewhere between true synaesthesia with visual image[1] and a more ambiguous complex of auditory, visual and emotional stimuli. We are concerned here not with how specific sources may evoke particular images but with how the imagery evoked interacts with more abstact aspects of musical composition.

In my discussion of music, I would like to use the term 'mimesis' to denote the imitation not only of nature but also of aspects of human culture not usually associated directly with musical material. Some aspects of mimesis are unconsciously passed on by a culture while others are consciously appropriated and used by the artist. Conscious and unconscious aspects are not sealed off from one another, of course, and a two-way exchange is evident over a period of time. We may have become much less conscious of the religious symbolism in Baroque music while being very conscious of the use of 'birdsong' in the music of Messiaen. There are two types of mimesis: 'timbral' mimesis is a direct imitation of the timbre ('colour') of the natural sound, while 'syntactic' mimesis may imitate the relationships between natural events; for example, the rhythms of speech may be 'orchestrated' in a variety of ways. In practice, from Janequin's *La Guerre* to Debussy's *La Mer*, the two types have been variously combined in what is known as 'programme music', as well as in the programmatic elements of much other music.

We must also be careful not to assume that the mimesis which might assist or motivate the composer is necessarily that which the listener will immediately apprehend. Trevor Wishart has argued that the greater the degree to which the composer has investigated the accepted mythic and symbolic structures of the culture of his potential audience, the greater this match will be – and arguably the greater the communication.[2] We will remain concerned here with the choices open to the composer of electroacoustic music, rather than the possible interpretation of those choices by the listener.

The use of natural sounds in the composition of electroacoustic music on tape allows us to claim that this is the first musical genre ever to place under the composer's control an acoustic palette as wide as that of the environment itself. Hence the vastly increased possibility that sounds may appear imitative. This contrasts strongly with the clear distinction, dominant in Western music aesthetics of recent centuries, between potentially 'musical' material based on periodic (pitched) sounds and 'non-musical' aperiodic sounds (noise). The evocation of image is further enhanced by a specific property of Western art: its deliberate removal from original context. Rarely does one view a landscape painting or listen to Beethoven's 'Pastoral' Symphony in a setting which is its apparent subject! By deliberately removing the visual clues as to the cause of sounds, indeed by removing or reducing visual stimulation of any kind, the composer is almost challenging the listener to re-create, if not an apparent cause, then at least an associated image to 'accompany' the music. The data for such a construction are entirely aural.

It is at this point that the composer must take into account audience response; he may intend the listener to forget or ignore the origins of the sounds used and yet fail in this aim. The earlier works of Pierre Schaeffer's group in Paris[3] (most notably Schaeffer's own *Etude aux Objets*) stubbornly refuse to relinquish this reference to the real world. The listener is confronted with two conflicting arguments: the more abstract musical discourse (intended by the composer) of interacting sounds and their patterns, and the almost cinematic stream of images of real objects being hit, scraped or otherwise set in motion. This duality is not new, as remarked above, and is familiar, for example, in the argument that Berlioz's *Symphonie Fantastique* is a better work than Beethoven's 'Battle' Symphony because the Berlioz has more 'abstract musical' substance, which is furthermore in finer balance with its programme – its mimetic content. The 'Battle' Symphony, like some early *musique concrète*, has been accused of being 'mere sound effects'. This 'abstract musical' substance I wish to redesignate 'aural discourse' to differentiate it clearly from 'mimetic discourse'. The two, to varying degrees in any specific work, combine to make the totality of 'musical discourse'.

For the composer of electroacoustic music this duality in content may be used to advantage. Even for those not interested in manipulating these associated images in composition, it must be at least taken into account. We must examine how these two possible approaches to language – which never exist in pure forms – might interact. Confining ourselves for the moment to works which deliberately use recorded sounds as material (not necessarily exclusively), we can see a continuum of possibilities between two poles. At one extreme, the mimetic discourse is evidently the dominant aspect of our perception of the work; at the other, our perception remains relatively free of any directly evoked image. From this continuum, let us draw for convenience three points of reference.

Works in which mimetic discourse is dominant include Luc Ferrari's series of works which he has described as 'anecdotal', including those entitled *Presque Rien*, and Trevor Wishart's *Red Bird*. In the Ferrari works, the composer uses extended recordings of environments: in *Presque Rien no. 1*, the sounds of the activities on a beach in the first few hours of the day; in *Presque Rien no. 2*, environmental sounds evoke a strange 'internal' travelogue. These recordings are left substantially unprocessed to 'tell stories'[4]. The Wishart work is subtitled 'A political prisoner's dream' and evokes images ranging from freedom to claustrophobic terror using human and environmental sounds. Works in which an aural discourse is dominant include many from composers based at the Groupe de Recherches Musicales in Paris in the period since the late 1950s. In many of these, while basic materials remained predominantly concrete in origin, the increasing sophistication of the possibilities of montage allowed a much more developed sound world to emerge, in which extended and complex sound-objects, free of associations, could be created. The earlier works of Luc Ferrari, Ivo Malec and François Bayle, and more recently Denis Smalley's *Pentes* may be included here. Between these two extremes lies our third reference point; an interesting balance of the two may be found most notably in such works as Bernard Parmegiani's

Dedans-Dehors, Michael McNabb's *Dreamsong* and Luigi Nono's *La Fabbrica Illuminata*. In all these works, from a diversity of traditions, the listener is aware that while recognition of the source of many of the sounds is intended, the impressions are welded together in other ways than those based on associative image.

This first series of examples is too crude, and I wish now to look at the term 'syntax', to analyse more carefully the possible approaches a composer may have to the organization of material, whether in aural or mimetic discourse.

Abstract and abstracted syntax

Much more analysis in recent years has been concerned with the relationships between objects or events and their possible transformations, rather than the nature of the events themselves. Lying behind such a structuralist approach is the premise that there exist universal forms of thought in any human mind, giving rise to specific utterances whose objects are defined by the particular cultural environment. We must examine how the failure of such a methodology to consider its use of terms such as 'law' ('rule') and 'explanation' has lead to significant ambiguities in terminology which have important consequences in discussion of contemporary music in general and electroacoustic music on tape in particular. Paradoxically, Claude Lévi-Strauss, while holding an important position in the structuralist pantheon, gives a quite clear pointer to this problem when he writes in the 'Overture' to *The Raw and the Cooked* of two levels of articulation of language:

> the first level of articulation . . . which consists precisely of general structures whose universality allows the encoding and decoding of individual messages . . . This is only possible because the elements, in addition to being drawn from nature, have already been systematised in the first level of articulation . . . In other words, the first level consists of real but unconscious relations which . . . are able to function without being known or correctly interpreted.[5]

The second level is that of the messages themselves. He goes on to imply that it is the absence of this first level – "unconscious relations . . . able to function without being known or correctly [i.e. consciously] interpreted" – which creates a fundamental problem both for what he describes as 'musique concrète' and for serial music.

It is true that the aims of both these kinds of music may be summarized as the discovery and use of 'universal laws'. These correspond to the 'general structures', referred to by Lévi-Strauss, which ideally form the basis of their communicability. However, the ambiguity in the term 'law' or 'rule' now emerges. The philosophers of science tell us of two traditional interpretations of these words[6]: law as an 'empirical generalization', that is, a summary of all observed instances of a particular event, and law as a 'causal necessity' having some sort of status 'above' the events and *determining* their occurrence. These interpretations are often confused in the arguments of musicologists and composers. The determinist and serial tradition tends to favour the latter interpretation, while (although some of the writings of Pierre Schaeffer appear to aim at a similarly universal 'solfège') the GRM group in Paris has developed a systematic approach favouring the former interpretation. Schaeffer's *Traité des objets musicaux*[7] is an attempt to establish rules for the combination of sounds, abstracted from an analysis of their perceived properties. This interdisciplinary approach is essentially empirical.

It must not be thought that the composer seeking this type of solution to musical organization starts entirely without preconceptions. Sound-objects do not suggest their own montage in an objective way! There lies, above the process of aural choice advocated in this approach, a set of beliefs as to what it is that 'sounds right' in any given situation. Loose terms such as 'gesture' may abound, but it is to this area, combining psychology of music with investigation of deeper levels of symbolic representation and communication, that future research must urgently be addressed. Such value systems remain to a large extent unconscious; we are not aware at the moment of perception why it is that a particular combination of sounds 'works', although we may rationalize our choice later and attempt a full explanation of what we have done.

It is only at this stage that we may examine the second of the two key words which the rationalist (and determinist) tradition of music composition has confused: 'explanation'. The perfect explanation of an event is so complete that it may be predicted[8]; for example a complete explanation as to why the sun has risen this morning and on previous mornings will allow us to predict with certainty that (given the same premises as pertained today) the sun will rise tomorrow; explanation and prediction are thus ideally symmetrical. While arguably true in the physical sciences, in the human sciences and the arts 'complete' or 'ideal' explanations are never attained, and the symmetry is flawed. It is not merely in post-war serial and other rationalist theories of pedagogy that analysis (explanation) of what has gone before is used as the basis for composition (prediction); it is in many ways the basis of the pedagogy of tonal music as well. Works are analysed, principles of an abstract nature deduced; works are then created from these principles (pastiche composition). Such a process may work reasonably well given an environment which encourages criticism and the serious reconsideration of its principles – in other words a simple feedback loop to enable the composer to modify his ideas. But one which fails to embrace this empirical and critical approach cannot hope to develop its language significantly.

This argument concludes that analysis and explanation should not be used blindly as the basis for systems or theories of composition, but can act only as pedagogic tools: part of a process of understanding, refining and perhaps assisting, decisions based on perceptual criteria. We may contrast this with traditions of contemporary composition whose essence is that of the creation and manipulation of essentially *a priori* shapes and structures by the composer. Serial composition is an important part of, but by no means alone in, this field. From the use of star maps to mystical number grids and formulas the use of principles not derived from the sound materials themselves all fall into this category. We may observe two quotations from Pierre Boulez to illustrate this point:

> It is important to choose a certain number of basic concepts having a direct relationship with the phenomenon of sound, and with that alone, and then to state postulates which must appear as simple logical relationships between these concepts, independent of the meaning attributed to them.[9]

These 'postulates' and their relationships in deductive systems have a status apparently autonomous from the particular musical material, although related to sound structures in general. Furthermore:

> . . . in relying almost entirely on the 'concrete, empirical or intuitive meaning' of the concepts used as starting points, we may be lead into fundamental errors of conception.[10]

While Boulez at least relates the origins of his basic concepts to sound, Stockhausen, during the 1960s, increasingly refined his ideas of abstract proportion derived from serial and numerical models with no apparent musical origin. We see the increasing use of the Fibonacci series to determine the section durations in *Mikrophonie I*, *Mikrophonie II* and *Telemusik*, apparently independently of the specific materials used. He finally reduced the definition of the musical objects to zero, defining only the abstract processes of transformation, as in the series of works using the 'plus–minus' notation[11], or even in some of the 'texts for intuitive music'[12]. In such ideas, Webern's 'nomos' – the 'law' immanent in the 12-note row – has finally been elevated to an abstract, near mystical, principle.

The abstract rules of Boulez and Stockhausen part company entirely at the formal level. Boulez rejects any referential or mimetic material, as such characteristics could never be "adapted to each new function which organises them";[13] he sees serial organization as pertaining directly to, because derived from, the parameters of a purely aural ('abstract musical') discourse. Stockhausen, however, in applying his abstract ideas to the formal levels of a work, allowed mimetic discourse to reappear; the two obvious cases are those of *Telemusik* and *Hymnen*, in which much is lost if the listener does not associate the folk or anthem materials with wider images of human culture.

So we may interpret the contemporary music polemic of the post-war era – in Europe the divide between 'elektronische Musik' and 'musique concrète', in America the divide between the legacy of serialism from Europe and the freer approach of many younger composers – as the opposition of an 'abstract' syntax to one 'abstracted' from the materials. In practice these two Utopian positions are rarely found in isolation, and many composers wander somewhat uneasily between the two.

* * *

The remarks above derive equally from considering examples whose primary concern – whatever the origins of the sound material – has been an aural discourse, and from those which – using the images evoked by referential material – make a conscious attempt at a mimetic discourse. For composers who have attempted to tackle this latter aspect, the same split between 'abstracted' and 'abstract' principles applies. The composer may preserve the relationships of the sound-objects made in an environmental recording, for example in the Ferrari works already referred to, thus 'abstracting' his syntax from them. Deliberately abstract forms and relationships may, however, be created as the basis of the montage, as in Stockhausen's *Telemusik*. Or the reconstructed image may be manipulated into unexpected juxta- and superpositions not usually encountered in the real world, creating surreal dreamscapes or dialectical oppositions, thus superimposing a 'story line' upon the material, and mediating between these two extremes. Trevor Wishart's *Red Bird* will be a case study for this approach.

Therefore, in summary, we may see the possible languages of electroacoustic music on tape in two dimensions. From one angle we may hear the music as having either an aural or a mimetic discourse; from another, either of these may be organized on ideas of syntax either abstracted from the materials or constructed independently from them in an abstract way. As we have noted, these are not fixed possibilities – in fact our one-dimensional line at the end of the previous section has become a two-dimensional plane over which the composer is free to roam. The next section will examine in more detail this 'language grid' (Fig. 1) and give examples of electroacoustic works to illustrate more clearly the problems and dilemmas which composers have encountered and their tentative solutions to them.

Case studies

In examining works to see how our discussion of approaches to language operates, we will consider, as in Figure 1, the degree to which the composer has decided to use mimetic reference as a compositional device, as the major axis. As in our first crude distinction, made at the end of the first section, we will divide this arbitrarily into three: I, works in which aural discourse is dominant; II, works in which aural and mimetic discourse combine; and III,

Abstract syntax	1	4	7
Combination of abstract and abstracted syntax	2	5	8
Abstracted syntax	3	6	9
	1: Aural discourse dominant	II: Combination of aural and mimetic discourse	III: Mimetic discourse dominant
	MUSICAL DISCOURSE		

Figure 1: Language grid

works in which mimetic discourse dominates. In each of these three categories, however, we will examine a further subdivision into three, along the second axis at right angles to the first, depending on the syntax type: works in which an abstract syntax has been imposed on the materials, works in which there appears to be a combination of abstract and abstracted aspects of the syntax, and works in which the syntax has been abstracted from the materials.

It cannot, however, be stressed too strongly that these nine 'compartments' are arbitrary subdivisions of a continuous plane of possibilities, the outermost boundaries of which are ideal states which are probably unobtainable.

I. Aural discourse dominant

Let us examine the types of electroacoustic material most likely to be found in works in which the composer intends a primarily aural discourse. While sounds of electronic origin tend to have less potential for mimetic reference and those of concrete origin potentially more, there is no such clear distinction in practice. Sounds of electronic origin may be used to evoke strong real-world images, as we see in Subotnick's *Wild Bull*, for example, or in Ligeti's *Artikulation*. In this latter work, while the composer's intention to model the electronic sound materials on speech was hardly an attempt to imitate but rather a simple means of differentiation, the sense of 'conversation' is encouraged through the use of statement–response, monologue, dialogue and 'argument' gestures. In this case syntactic mimesis predominates over timbral mimesis. Conversely, the development of sophisticated tape techniques in the 1960s lead to a genre of works in which concrete sounds could be manipulated as abstract objects depending on their acoustic content, losing all reference to their origins or environment. These sounds appear increasingly 'electronic' – though in a particularly complex sense – and lend themselves ideally to being manipulated in a purely aural discourse. The listener need not know the origins of the material which, whatever its electroacoustic origins, lacks reference to images of the real world.

The most obvious types of material least likely to invoke images of or references to the 'real' world are those sounds of electronic origin not immediately modelled on sounds of the environment. The two *Electronic Studies* of Stockhausen from 1953–4 are examples in which the materials, created entirely from sine waves in combination, avoid any attempt at imitation not only of environmental sound, but even of instrumental timbres.

Most electronic synthesis is modelled on instrumental or vocal sounds and thus at a deep level few sounds are entirely free of all mimetic reference. We may concede that instrumental sound-images and evocation, being primarily musical, may still be allowed within this category. It is the sounds of the environment not traditionally associated with music whose imagery we wish to discuss as mimetic discourse. Thus Stockhausen's *Kontakte*, while containing an enormous variety of quasi-instrumental sounds, remains primarily 'aural' in all aspects of its discourse.

The use of speech in electroacoustic music presents this categorization with a problem. Yet another layer of articulation is apparent in that speech may evoke images due on the one hand to its acoustic nature and on the other to its actual meaning. In order not to confuse the issue, let us consider that speech used primarily for its phonetic content is being used in an aural way, while that used for its semantic content (including gestural contour) contributes to more mimetic possibilities. Let us now look at the methods that composers have used in the organization of these non-referential materials into musical form. As outlined above we will examine the range from *abstract* to *abstracted* syntax.

1. *Aural discourse: Abstract syntax.* Applications of serial principles to electronic sources give us the clearest cases in this first category. The principles of organization of works as diverse as Milton Babbitt's *Ensembles for Synthesizer* or Stockhausen's *Electronic Studies* concern the creation and manipulation of abstract shapes created independently of the *perceptual* qualities of the materials used. Of course, the particular serial patterns chosen may be influenced by the materials of the work. In the case of Stockhausen's *Study II*, the composer, in his search for a principle of unity to determine both the organization and the acoustic properties of the sounds, used the same numerical ideas. The individual components of the 'tone mixtures' are based on the same interval system as that used to organize the mixtures themselves.[14] This is not the same as an attempt to draw language from the intrinsic nature of the sounds. While an attempt has been made to integrate and relate the two aspects – material and structure – the syntax still originates from an abstract domain, superimposed on, and not drawn from, perception of the sounds themselves. In *Kontakte* Stockhausen moved one small step away from such slavish adherence to the ideal. Heikinheimo has pointed out[15] the interesting discrepancies between the composer's intended section ('moment') lengths and what actually resulted in the studio, and it is evident that Stockhausen is making many more individual decisions based on the

evidence of the ear. *Kontakte* is thus somewhat further along our scale towards the *abstracted* pole.

A surprising group of works to find in the same company as these serial or determinist examples are those based on an apparently diametrically opposite philosophy, but one equally remote from perceptual properties of the musical materials. Those works of John Cage which utilize chance procedures, such as *Williams Mix* or electroacoustic realizations of *Fontana Mix*, all apply abstract schemata in one medium or another to the creation and montage of sound materials. *Williams Mix* is a tape composition in which the composer, after division of the material into seven categories (only one of which covers sounds of exclusively electronic origin), used *I Ching*-based chance procedures to determine the details of the montage.[16] The score of *Fontana Mix* consists of a series of plastic strips whose disposition after being let fall may be used to determine both the materials and organization – created in performance or fixed in advance – of any number of versions.[17] It is not the origin of these schemata which is at issue, but their existence prior to the perceptual properties of the particular materials which they create or organize.

2. Aural discourse: Combination of abstract and abstracted syntax. The next group of works to consider within this section are those which harness some aspects of the perception of sound, and yet rely on formal schemes at other levels of language. The composer may have conducted extensive analyses of the materials and then integrated this empirical evidence within shapes and patterns of a more abstract nature. This has most often emerged in the works of composers who have attempted to apply the rigour of serial thinking to more perceptual aspects of sound than the crude parameter applications of the 1950s. Stockhausen appeared to be moving to this position in such non-electroacoustic works as *Momente*, in which scales based on equal perceptual intervals were constructed; it is interesting to note the influence of electroacoustic thinking on this work in that the primary scale elaborated is that from 'note' to 'noise'. Electroacoustic composers of that era still lacked the technical means to investigate and re-create such a range of possible timbres, and controlled experiments had to wait until sophisticated computer-aided analysis and resynthesis became available in the 1970s.

Jonathan Harvey's *Mortuous Plango, Vivos Voco* is a case in point. Composed at IRCAM[18] in Paris in 1980, the work combines use of the Stanford[19] sound analysis package with MUSIC V[20] and the more recent CHANT program[21]. The composer based the pitch series, used as the central tones of each of the eight sections of the work, on partial tones drawn from a spectrum analysis of the great tenor bell at Winchester Cathedral; many other aspects of its material and organization derive from this series, including the ways in which recordings of a chorister's voice are tuned and combined to blend with the bell sound. Such a process could have been applied in an entirely abstract manner, but Harvey has sensitively combined these schemes with striking integrations and transformations of timbre, carefully controlled and modified by aural perception of the results.[22]

3. *Aural discourse: Abstracted syntax.* Our third group within this category of aural discourse consists of those works whose composers have attempted to construct a syntax abstracted from perception of the material itself. While this was the avowed aim of the GRM from its inception in 1948, its first major successes did not appear until the late 1950s and early 1960s; from the reorganization of the group in 1957, the sound-object could, through far more extensive manipulation and investigation, be made completely independent of any reference to source. The real possibility of a unity of the sound worlds of *musique concrète* and electronic music existed for the first time. They were, however, to remain stubbornly separate in their approach to organization of this material, and the advent of voltage-controlled electronic synthesis in the 1960s was, in the short term at least, to perpetuate this divide. The approach examined here reached the highest levels of technical and musical refinement in the 1970s.

Most of the materials for Denis Smalley's *Pentes* ('Slopes') are derived from the sounds of small percussion instruments and the Northumbrian pipes. Through a montage process involving innumerable juxtapositions and mixing routines, the composer has created textures and drones, complex attacks and continuants, assembled in a way which is clearly dictated by aural judgments of gesture and patterns of growth. The overall harmonic plan is minimal and does not function as a carrier of interest; in the composer's words:

> A significant feature of *Pentes* is the slow evolution of a harmonic progression introducing the Northumbrian pipes' melody. If played on the piano this progression would appear mundane. However, . . . its temporal elongation and the careful revelation and control of the internal, fluctuating harmonics extracted through transformations in the studio ensure that many more qualities contribute to its impact than merely the effect of the chord progression alone. Interest focuses on the subtle pulsed shifts in the harmonic spectrum.[23]

With the exception of the pipes' melody referred to, the origin of the materials is entirely and intentionally lost in a sound world of enormous subtlety and power. This approach is summed up in the credo:

> Today it is ironically necessary to reassert the primacy of the aural experience in music; perceptual acuity and experience are often more reliable and valuable in this search than formal research.[24]

This effectively reiterates two principles of research into the sound-objects of *musique concrète* – 'Primacy to the ear!' and 'Search for a language!' – first propounded by Pierre Schaeffer's Paris group in 1953.[25] *Pentes* was commissioned by the GRM and composed in its studios in 1974.

Bernard Parmegiani's *De Natura Sonorum* (1974–5) is from the same era and studio. The work is in ten independent movements, which the composer has grouped into two series (with some variants in performance), each an 'étude' concentrating on one aspect of the sound-object or a principle of montage. This is the form most favoured by Parmegiani in works in which he intends to minimize mimetic reference. In this case sounds of electronic origin are combined with those – mostly instrumental – whose origin we can, if we wish, reconstruct while listening. But in this case the composer skilfully combines the material in ways which concentrate on the perception of specific acoustic properties, moving our attention away from any possible mimetic references, not merely towards the microstructure of the sounds but towards the way the sounds combine to reinforce this perception. This is most striking in the movement 'Incidences/Résonances' (the first in the first series), in which the attacks and resonances of natural instrumental sounds are combined with – and effectively prolonged by – sounds of electronic origin. This is not simply 'colour composition', but a process in which dynamic gestures and organic structures evolve and form an important part of the language, which remains rooted in the nature of sound itself.

II. Combination of aural and mimetic discourse

We will now consider three works or groups of works which attempt to combine an aural discourse with a mimetic one; in other words the composer intends the listener not only to appreciate the more abstract aspects of the work, but also to recognize and appreciate a series of images evoked by the material as an integral part of the composition. This dual use of material may, as before, be structured by a syntax varying from the abstract to that abstracted from these materials.

4. *Combination of aural and mimetic discourse: Abstract syntax.* Luigi Nono was one of the original members of a group of European composers who have come to be known as the 'Darmstadt school' owing to their association with the summer school of that name in Germany in the early 1950s. Nono's political commitment has, until recently, been explicit in his music; it is not surprising, therefore, that in his avowed advocacy of a left-wing, 'realist' view of art, he should seek to convey his strong views on this reality in the use of explicit texts and images. Between 1950 and 1960, Nono's works were entirely instrumental and vocal, developing and extending his ideas of serial technique. Then, quite abruptly in 1960, an invitation to the electronic music studio of RAI Milan[26] resulted in his discovery of the potential of the tape medium for these ideas.

The dialectic is the basis of much Marxist discourse. Two opposites are juxtaposed and form a new relationship (thesis, antithesis, synthesis), this relationship in turn creating a further dialectic, and so on. Nono's music may be seen as applying this process on several levels. In many of his works involving tape, the ending is often musically unsatisfying, abrupt and unexpected or faded out, suggesting, perhaps, his belief that a final synthesis is not possible – the struggle is not complete. He juxtaposes apparently irreconcilable materials. In *La Fabbrica Illuminata* (1964) for soprano and tape, the most important dialectically opposed pairs are 'man:machine' and 'individual:group'. For Nono, the voice is the essentially human agent, pitted on the one hand against factory sounds, on the other against crowd sounds, all intended to relate to the experience of the audience and to provoke discussion. In addition to this 'programme' behind the music, Nono retains a determinist (usually serial) approach to much of the work's organization, in the ordering both of the pitched material (for the soprano, both live and pre-recorded) and of the 'blocks' of factory sounds. The pairs of opposites overlap and develop. Their interrelation may be shown in a simple model, Figure 2.

It is evident that the tape carries the most developed parts of the argument, and itself embodies the original 'man:machine' division on which the whole work is based. In *La Fabbrica Illuminata* there is no mediation between these extremes. In this respect Nono's work is an example of a true collage principle, in sharp contrast to the prevalent ideas of that era of transformation and mediation functions derived from aspects of serial principles[27]. The

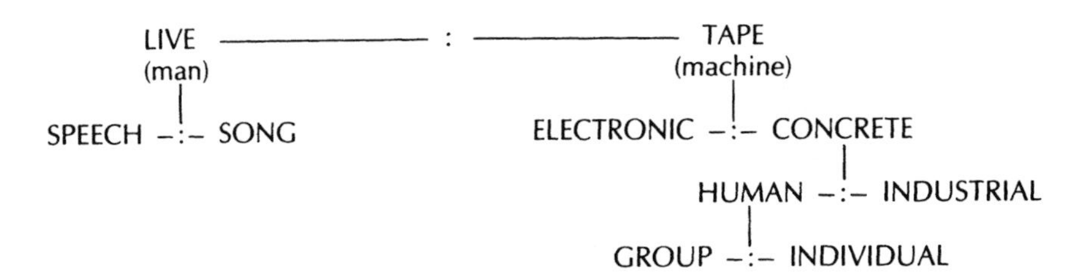

Figure 2: Dialectical opposites in the music of Luigi Nono

'tree-like' model of dialectical pairs above cannot, therefore, be interpreted in the same way as, say, the form plan of the materials of Stockhausen's *Momente* in which similar binary opposites are repeatedly divided[28]. Some of Stockhausen's other proportion systems define a mediating role and a continuum between any two of the sound categories. For Nono, the incompatibility of lyricism and the violence of exploitation cannot be mediated, which is why, at first hearing, so many of his works appear unintegrated and inconclusive.

It is thus not so much the acoustic nature of the recorded sounds which defines the syntax of the work – Nono could not have accepted such an 'aesthetic' position at this time – but the nature of the origins of those sounds in society which he seeks to illustrate. This therefore remains an abstract syntax in that it is one not derived from acoustic properties of the sound materials, albeit one intended to have a very explicit narrative function.

5. *Combination of aural and mimetic discourse: Combination of abstract and abstracted syntax*. The next area to consider in our language grid lies in the centre at the balance point both of discourse type – aural and mimetic – and of syntax type – abstract and abstracted.

Michael McNabb has written concerning his work *Dreamsong* (1978):

> The basic intent of the piece was to integrate a set of synthesised sounds with a set of digitally recorded sounds to such a degree that they would form a continuum of available sound material. The sounds thus range from easily recognisable to the totally new, or, more poetically, from the real world to the dream world of the imagination . . .[29]

The listening experience confirms the composer's intention. There is a fine balance in the combination of mimetic discourse – a series of dream images and transformations of 'real' objects – with more aural ('abstract musical') organization of pitch structures and timbres.

In some cases the composer has re-created apparently real sounds using synthesis techniques: a cluster of bell-like sounds coalesces into a vocal sound, a chorus of voices descends into a bass drone. Such seamless transformations are still easiest using additive synthesis techniques to allow complex interpolation between the spectrum types. The data for this synthesis were based on Fourier analysis of the original recorded sound.[30] The development of this 'analysis–resynthesis' approach at Stanford (subsequently unitlized at IRCAM) is the fundamental tool which will allow the integration of electronic and *concrète* techniques. Even the most complex pre-recorded sound used, that of a crowd (perhaps awaiting a performance of some kind), is phased and filtered in such a way as to pass easily into the pitched material which follows. The fact that the synthesized sounds used are almost instrumental in nature (rather than complex timbres or textures) allows one to see the work in terms of a 'dream orchestra' in which all sounds have become instruments and in which one instrument may be transformed into another.

This approach to timbre makes demands on the syntax of the work in terms of combinations of sound-objects and their possible durations and structures. However, the composer has united these limited demands with a more abstract and traditional conception of technique. The work is based on two modes: one myxolidian based primarily on B flat, the other a synthetic mode of semitones, tones and thirds ranging over two octaves. In addition, two themes are drawn from these modes, one of them being a fragment of a Zen sutra.

On matters of duration McNabb writes:

> . . . most of the slower rhythms and section lengths derive from Fibonacci relationships, not because of their numerologic or mystic implications, but because they present a convenient and effective alternative to traditional rhythmic structures. Of course, a little acknowledgement of the gods of mathematics never hurt any computer musician.[31]

This is an interesting admission to a pragmatic approach to systems. There is no doubt that the composer would have tried another approach had the Fibonacci series resulted in unsatisfactory aural results; it is up to others interested in the application of artificial intelligence to future programming to investigate under what conditions such systems might 'work'.

The fact that this work sits exactly at the fulcrum of our language grid, balancing so many of the tensions of materials and technique that had emerged over the previous thirty years, should not in itself give special credence to the ideas that McNabb has put forward. But it is precisely his success in uniting and transcending these disparate forces which suggests that *Dreamsong is* a pivotal work, an essential landmark for the fourth decade of electroacoustic music.

6. *Combination of aural and mimetic discourse: Abstracted syntax.* Finally in this group of works we must examine an example of a tape piece which, while combining aural and mimetic discourse, seeks to derive its syntax entirely from the acoustic properties of the materials themselves.

In all of Bernard Parmegiani's works a very special role is played by the natural sounds he has recorded. His is the most acute ear for spectral detail; each sound is recorded in such a way that its internal structure is the object of intensified perception. He has succeeded in creating an aural discourse from the subtleties of these sound-objects while still allowing, and in some cases encouraging, recognition of the source of the sounds. He has achieved in several works a simultaneous exposition of aural and mimetic structures in which the two interact and support each other to such an extent that at most times no distinction can be made. This is the case with *Dedans-Dehors*, composed in the GRM studios in 1976–7. The composer's note to the work[32] encapsulates just this balance. On the one hand he explains some of the purely technical aspects of its syntax: the principles of metamorphosis of the sounds and their morphologies. Yet he goes on to list ten 'sound symbols' related to the basic idea of 'within-without' – some elemental or natural, pertaining to earth, air, fire, water and animal sounds, others human-created. The work, while playing continuously, is divided into ten sections: 'En phase/hors phase', 'Jeux d'énergie externe et interne', 'Retour de la forêt', 'Action éphémère', 'Métamorphose 1', 'Métamorphose 2', 'Le lointain proche 1', 'Le lointain proche 2', 'L'individu collectif' and 'Rappel au silence'. While aural and mimetic references are mixed in these titles, there is complete balance of the two in a structure whose order is drawn entirely from the perceived nature of the materials: the juxtaposition of long drone sounds and exquisitely recorded fire and water sounds, for example, has just the right sense of timing and spectral contrast related to feelings of human and natural time-scale and gesture.

It could be argued that the language of *Dedans-Dehors* in its references to obvious narrative and extra-musical images may move somewhat towards the abstract pole. We must be clear about exactly what aspects of a work are pre-defined by having such a narrative. In the case of Parmegiani, the composer uses narrative image ideas at a stage in the composition where he is building up a basic library of sounds. The images suggest the sound world. Unlike the works to be discussed in the following section, he does not then construct the finished work with a view to communicating this image sequence directly to his audience. The narrative remains, but in fine balance with a sense of aural flow which is the result of a montage in which the moment-by-moment combination of sounds is judged and adjusted by aural criteria. We may see a parallel in earlier music collage forms: the programme of Stravinsky's *Petrushka*, for example, will influence its structure and the types of materials used in each block of sound, but the note-by-note composition within these formal constraints is not defined. Such preconceived narrative image structures do not therefore *necessarily* interfere with choices and judgments of sound combinations – though they may be made to do so if applied too literally.

III. Mimetic discourse dominant

The return to an interest in mimetic reference in electroacoustic music, at least in Europe, came as a reaction against the developing sophistication of tape, and hence sound-object, manipulation of the late 1950s and early 1960s.

Stockhausen's *Telemusik*, Trevor Wishart's *Red Bird* and Luc Ferrari's *Presque Rien no. 1* have been chosen to illustrate the abstract/abstracted distinctions in this group. They have much in common. All have aims apparently outside those traditionally accepted as 'musical': the Wishart and Ferrari, overtly in terms of political or social issues, the Stockhausen in terms of an attempt to integrate many disparate musics of the world. There is interestingly a little-discussed connection between the ideas of Luc Ferrari and Stockhausen in the period 1963–6.

Luc Ferrari worked at the GRM in Paris from 1958 to 1963. His first works were in the then prevalent 'abstract expressionist' genre, in which the sounds were hardly ever recognizable, and the principles of organization were basically those at the heart of Schaeffer's thinking. In the early 1960s he felt it necessary to return to an approach in which the sounds could be exploited with respect also to a recognition of their origins – 'electroacoustic nature photographs'[33]. Ferrari termed the reintroduction of this narrative element into the materials 'anecdotal music'. His first work in this genre was *Hétérozygote*, composed in 1963–4. While we shall return to a discussion of this genre with respect to a later work (*Presque Rien no. 1*), this work proves to have important links to Stockhausen's *Telemusik* and the later *Hymnen*.

Ferrari left the GRM in 1963 and was invited by Stockhausen, as director of the Cologne New Music Courses, to be course director for 'musique concrète' in the 1964–5 and the 1965–6 courses. At the same time, Stockhausen had chosen Ferrari's earlier work *Tautologos II* as one of the music examples in the very first programme in the first of two Westdeutscher Rundfunk radio series entitled 'Do you know music that can only be heard over loudspeakers?', which he presented between 1964 and 1966. The same work was also presented during the second Cologne course in December 1965. *Tautologos II* combines a vast variety of sound types, some recognizable, some ambiguous, some abstract. Furthermore, Ferrari was at work on *Hétérozygote*, his first 'anecdotal' composition, at just this period. Early in 1966, Stockhausen left for his first visit to Japan, during which *Telemusik* was composed. On his return in April of that year, Stockhausen embarked on the second series of broadcasts, not merely including Ferrari's *Hétérozygote* in his review of the work of the GRM, but devoting considerable time to discussing it:

> Such a mixture of nameable and nameless, defined and ambiguous sound events jump to and fro between outer, objective situations, and the inner imaginary subjective sound world . . . It appears to me that the music of the immediate future will essentially be determined from such relations . . . linking musical photography (by that I mean exact reproduction of acoustic events) with free sound images . . . *Hétérozygote* . . . [is] remarkably independent, open, plural . . . Discovery of the subtle rules of relation of this new polyphony will be the task of the immediate future . . .[34]

It should be noted that for Stockhausen the "discovery of the subtle rules of relation" was rarely empirical and should more correctly be termed 'invention', though there is much greater freedom in the language of *Hymnen*, the work which immediately followed the broadcast, than in that of *Telemusik* which preceded it.[35]

7. *Mimetic discourse: Abstract syntax*. The structural plan of *Telemusik* uses the proportions of the harmonic series (what Stockhausen terms a

'formant rhythm') in which each 'node' of a 'harmonic' is articulated by an attack on a (pre-recorded) Japanese temple instrument – a different instrument for each harmonic. This process determines the basic order of the 32 sections of the work. As all but prime numbers will produce coincidences in this scheme, the composer modifies the details to avoid these. The sections generated from the same harmonic are all assigned a specific duration derived from the Fibonacci series. Thus the function of the instrumental strokes is such that each instrument indicates the duration of the succeeding section. The composer also modifies the number of occurrences of each section type to fall on a Fibonacci series value. Disregarding slight discrepancies in the resultant lengths, there is one section of 144 seconds, there are two of 89 seconds, three of 55 seconds, eight of 21 seconds, and thirteen sections of 13 seconds.

Each moment was designed to be realized with relative independence from the others in the short time available. Stockhausen created an abstract architectonic form into which the material was 'poured'. While the mimetic nature of the sounds – pre-recorded examples of folk music from around the world, with additional Japanese material recorded specially – is obscured by the electronic process of ring modulation, the vision of a 'radiophonic' integration of all such types of music remains very powerful.

The composer has stated:

> Metacollage means . . . going beyond collage . . . Collage is glueing together and seeing what happens. It's not really mediation . . . Our music represents models of elements that are very heterogeneous and seemingly unmatchable . . . These are complementary societies and structures . . .[36]
>
> You hear this 'found music' not merely as it was originally, but I have tried to bring those apparently so heterogeneous appearances into a close relationship.[37]

Thus the motivation towards the material ('found music') lies with respect to its origins and associations, which the composer seeks to 'intermodulate', both literally – the superimposition of the rhythm of one on the timbre of another – and spiritually, into a 'world music'.

8. Mimetic discourse: Combination of abstract and abstracted syntax. The symbolic representations underlying the mimetic references in Trevor Wishart's *Red Bird* are discussed fully in the score[38]. It represents a work in which mimetic discourse is dominant and whose syntax combines montage based on both the specific acoustic properties of the sounds and a more abstract schema based on a carefully determined symbolic narrative. It was argued in the previous section that narrative ideas used to determine the form of a work did not necessarily negate the idea of a montage based on sound relationships alone. In the case of Parmegiani's *Dedans-Dehors*, such a narrative established the boundaries of the sound world and the library of sounds the composer wished to establish. This is true of *Red Bird* also, but the function of the narrative is substantially more determinate. The final order and combination of sound-events is strongly influenced by what is effectively a 'story line', while the composer retains an aural judgment as to the exact nature of many of the studio montage procedures. The work thus combines elements from 'abstract' and 'abstracted' syntax poles.

At all times the sound materials refer to images of the real world; Wishart's commitment to a realist philosophy parallels that of Luigi Nono, in that they both see music as an active ingredient in the formation and reformation of consciousness, but differs from it in its insistence that renewal of language must be based on a corpus of existing and widely held ideas and symbols. Wishart develops (where Nono does not) the idea of image transformation: words 'become' birds, clock ticks 'become' the slamming of the prison door. As in the works of Parmegiani, this juxtaposition produces a surreal edge to the realist associations: not the egocentric self-indulgence of many surrealist images, but a violent and alienated nightmare which has all too real an existence.

> . . . the 'drama' is played out entirely in the transformation of sound elements of a symbolic landscape . . . Even where the sounds are at their most abstracted [see Note 39], they can always be related back to the recognisable sound symbols out of which they have been evolved by transformational procedures.[39]

It is, from the composer's point of view, a balance of abstract and abstracted composition syntax, with a bias towards the latter. Having established the idea of a myth-narrative structure based on symbolic sound references, Wishart goes on:

> What is needed is some means of systematically forcing the imagination to consider the possibilities from as many different points of view as possible. This was achieved by using a

> permutation procedure akin to those used by serial composers, but different in one fundamental aspect, ie. this is an OPEN-ENDED procedure, a heuristic tool to define a method for thinking about possibilities, and is in no way definitive of a set of SPECIFIC musical materials, or a mode of organisation.[40]

Wishart continually stresses the organized nature of procedures of discovery and subsequent cataloguing into morphologies and typologies. Thus the overall myth plan defines only the broad image transformations, the actual sound-manipulation procedures are a matter for studio experiment and discovery, based on aural judgment. *Red Bird* remains one of only a handful of works to have grasped mimetic transformation as a vehicle for the communication of ideas.

9. *Mimetic discourse: Abstracted syntax*. Finally, in our discussion of all the possible combinations of discourse and syntax type we must consider the possibility of a work in which mimetic discourse is dominant and a syntax abstracted from the materials is developed. At its purest this might be represented by an environmental recording minimally edited or altered. There may be reasons for certain choices – location, time of day, duration – which remove the work from the entirely arbitrary. Luc Ferrari's *Presque Rien no. 1*, composed in 1970 and subtitled 'Daybreak on the beach', is an example of 'anecdotal music'. It consists of a straightforward recording of several hours duration of the activities on a beach around sunrise. This has been judiciously edited down to about 20 minutes duration without any noticeable discontinuities resulting. The composer has refrained from imposing any moment-to-moment montage on the materials and has described it as "more reproduction than production"[41]. If *Presque Rien no. 1* is 'minimal' electroacoustic music, not all Ferrari's anecdotal works are so entirely free of the composer's manipulations. The earlier *Music Promenade* (1968) is a polyphonic mix of several such layers with the heightened reality that this simultaneity of different environments brings about. This focussing and framing process using narrative natural sound sources, while respecting the autonomy of the original sounds, may be used therefore not to obscure but to heighten our awareness of the environment. The photograph is a good parallel in that it is so clearly not the original object itself, the act of 'recording' becoming part of the new artefact. The will of the composer, far from abdicated, is crucial.

* * *

We have now completed our zig-zag walk across the grid of syntax possibilities which we defined at the onset. All the approaches had in common the conscious involvement of the composer in the creation of musical syntax, abstract or abstracted. In comparing Ferrari's *Presque Rien no. 1* with Stockhausen's *Study II* at diametrically opposite corners of our grid, we may see the enormous distance traversed.

Conclusion

This discussion of the grid of possible discourse and syntax types for electroacoustic music on tape refers primarily to those works in which timbre ('colour') composition plays an important part. There remain genres of electronic and computer music which retain an 'instrumental' emphasis on pitch relationships. Almost all pitch-oriented electroacoustic music belongs in the first area we examined: the discourse is exclusively aural ('abstract musical'), the syntax almost always entirely abstract (often serial at root), not based on intrinsic sound-object relations. These have not been the main concern of this chapter.

Such a grid as has been presented here allows us to see more clearly differences between what are too loosely described in the literature as 'musique concrète', originating in Paris, and 'elektronische Musik', originating in Cologne. It is important to note that the introduction of electronic generators to the GRM studios in Paris, or of concrete resources to the WDR studios in Cologne[42], made little or no difference to the respective approaches to composition with which they are associated. It is a gross simplification to imply that Stockhausen's *Gesang der Jünglinge*, in using the recording of a boy's voice as part of the material, broke the barriers between the two groups.[43] The differences between the two approaches were fundamentally between the abstract and abstracted approaches to syntax. It is perhaps only with the developments represented in this discussion by Michael McNabb's *Dreamsong* that unity of both the aural/mimetic and the abstract/abstracted dimensions of the language of electroacoustic music becomes possible.

3

Sound Symbols and Landscapes[1]

Trevor Wishart

Chapter 4 of Pierre Schaeffer's *Traité des objets musicaux*[2] is entitled 'The Acousmatic'. According to the definition in *Larousse*, the Acousmatics were initiates in the Pythagorean brotherhood, who were required to listen, in silence, to lectures delivered from behind a curtain so that the lecturer could not be seen. The adjective 'acousmatic' thus refers to the apprehension of a sound without relation to its source. It was important in Schaeffer's development of the concept of the 'sound-object' that it be detached from any association with its source or cause. The sound-object was to be analysed for its intrinsic acoustic properties and not in relation to the instrument or physical cause which brought it into being. However, in our common experience, we are more often aware of the source of a sound than not, and studies of behaviour and aural physiology would suggest that our mental apparatus is predisposed to allocate sounds to their sources. We can see in a very crude way how this ability was essential for our survival in the period before our species came to dominate the entire planet. Even in the cultured detachment of today's world, however, when we are listening to a concert of instrumental music, except where the texture is very dense or produced in a way which is novel to our ears, we are always very aware of the instrumental source of the sounds we hear. The formalization of musical parameters – the lattice of the tempered scale, the rhythmic coordination required by harmonic structuration, the subordination of timbre to pitch and its streaming to separate instrumental layers – is in many ways an attempt to negate the impact of the recognition of the source (human beings articulating mechanical sound sources) and focus our attention upon the logic of the musical structures. Part of our enjoyment of the music, however, remains an appreciation of the human source of the sounds themselves, which is also in a sense distinct from the articulation of non-notated parameters of the sound through performance gesture.

In some contemporary instrumental music such as Boulez's *Structures*, or the early, purely instrumental, serial works of Milton Babbitt, it has been possible for a composer to specify a type of architecture and a mode of sound production which limits the possible impact of gestural characteristics upon the acoustic result, and when such music is heard on loudspeakers a large degree of detachment from recognition of the source may be achieved for some listeners. With music for voice, however, it is doubtful if we can ever banish from our apprehension of the sound the recognition of a human source for those sounds. Furthermore, that recognition often plays a significant role in our perception of the music itself. For example, in the *Trois Poèmes d'Henri Michaux* of Lutosławski, our perception of a mass of human 'utterers' is important in itself and becomes especially so in the crowd-like sequence, the recognition of 'crowd' or 'mob' contributing significantly to our aesthetic appreciation of the work.

At this stage, let us place these various characteristics of the sound experience related to our recognition of the source of the sounds under the general heading of 'Landscape'. It is important at this stage to differentiate the idea of 'landscape' from that of 'association' as it is frequently used with reference to programmatic music. Thus in our listening to the final movement of Tchaikovsky's *Manfred* symphony we may be led (by the programme note or otherwise) to associate the acoustic events with the idea or image of the distraught Manfred wandering through the forest; the landscape of Tchaikovsky's *Manfred* symphony is, however, musicians-playing-instruments. Landscape, at least at this level, is then a reasonably objective criterion, related to our recognition of the source of the sounds.

With the arrival of sound recording the question of source-identification of sounds increased in importance. Previously the landscape of a sound had been perceived as the physical source of the sound; what now was to be made of a recording of Beethoven's 'Pastoral' symphony played on loudspeakers? The physical source of the sounds is the vibration of the cones of the loudspeakers, but as the loudspeaker is able to reproduce sounds from any other source this tells us almost nothing about that sound except that it has been recorded. We must therefore seek a redefinition of the term 'landscape'. If the term is to have any significance in electroacoustic music we must define it as the source from which we *imagine* the sounds to come. The loudspeaker has, in effect, allowed us to set up a virtual acoustic space into which we may project an image of any real existing acoustic space, and the existence of this virtual acoustic space presents us with new creative possibilities.

The first group of composers to attempt to make music with pre-recorded sounds (*musique concrète*) was the Groupe de Recherches Musicales in Paris, whose most prominent member was Pierre Schaeffer. Although some of the earliest work of the group (e.g. *Symphonie Pour un Homme Seul*) was clearly dependent on the listener's recognition of the source of the sound material, this approach was quickly rejected. The philosophy of composing which gradually emerged, particularly from Schaeffer's writings, centred on the notion of the acousmatic and the abstraction of the 'sound-object' from any dependent relationship to its origins. One of the first composers to take really seriously the possibilities of landscape as a musical tool was Luc Ferrari:

> I thought it had to be possible to retain absolutely the structural qualities of the old musique concrète without throwing out the reality content of the material which it had originally. It had to be possible to make music and to bring into relation together the shreds of reality in order to tell stories.[3]

Ferrari described his approach as 'anecdotal'. In *Presque Rien no. 1*, musically the most radical work in this genre, he takes a recording of several hours of activity on a beach and compresses this material into 20 minutes, without in any way attempting to negate or transform (other than enhance by concentration) the perceived landscape of the beach. Parmegiani, on the other hand, developed a kind of surrealistic approach to recognizable sources which lay more easily within the *musique concrète* frame of reference.

Thus changes in aural perspective on an object can be obtained by recording it at a normal listening distance and then close-miking it. These approaches produce quite different acoustic results and when they are juxtaposed in the aural landscape our sense of aural perspective is transformed. Similarly, unrelated images such as water and fire, or clearly synthetic and clearly recognizable sounds, may be brought into close proximity and, by the use of similar musical gestures or the creation of similar overtone structures, made to have a very strong aural interrelationship.

A similar concern with aural landscape developed in the radiophonic departments of radio stations broadcasting drama. Whereas Parmegiani was concerned to hint at possible landscapes in an essentially acousmatic approach to sound materials, radiophonics was, at least initially, concerned exclusively with the suggestion of real landscapes which were not physically present in the recording studio. In most cases, these landscapes were associated with spoken text or dialogue which itself suggested a context. Often simple devices like distance from the microphone, the use of echo, or the sound of a door opening or closing might be sufficient to 'set the scene', but as time went on more elaborate scenarios were developed, from seventeenth-century battles to the purely imaginary landscape of alien spacecraft. The appearance of science fiction drama encouraged the development of less obvious landscape devices, but radiophonics largely confined itself simply to exploiting the unusualness or novelty of a sound-object (which very rapidly became clichéd and predictable). Before the manipulation of aural landscapes (see below) could be explored and developed, the concreteness of the television image took over and radiophonics declined to an even more marginal position.

Defining characteristics of landscape

The landscape of a sound-image we have therefore defined as the imagined source of the perceived sounds. The landscape of the sounds heard at an orchestral concert is musicians-playing-instruments. The landscape of the same concert heard over loudspeakers is also musicians-playing-instruments. In some cases it is difficult to identify the source of the sounds and this fact is particularly pertinent to music designed for projection over loudspeakers. When listeners accustomed to concerts where instrumentalists perform attend concerts of electroacoustic music projected on loudspeakers, they often express a sense of disorientation. This is usually attributed to the lack of any visual focus at the concert. However, it seems clear that this reaction is prompted by an inability to define an imaginable source, in the sense of a landscape, for the sounds perceived. This sense of disorientation produced in some listeners by the impact of electronic sounds was the basis of the early use of electronic sound materials for science fiction productions. The inability of the listener to locate the landscape of the sounds provided the disorientation and sense of strangeness which the producer wished to achieve. The development of the concept of the acousmatic and the general tendency in mainstream *musique concrète* to destroy clues as to the source or origin of the sounds can be seen as a specific reaction to the problem of landscape in electroacoustic music.

What aspects of our perception of an aural image enter into our definition of an aural landscape? We may effectively break down our perception of landscape into three components which are not, however, entirely independent of one another. These are: I, the nature of the perceived acoustic space; II, the disposition of sound-objects within the space; and III, the recognition of individual sound-objects.

I. The nature of the perceived acoustic space

Usually, any sort of live recording will carry with it information about the overall acoustic properties of the environment in which it is recorded. These might include the particular resonances or reverberation time of a specifically designed auditorium or the differences between moorland (lack of echo or reverberation, sense of great distance, indicated by sounds of very low amplitude with loss of high-frequency components etc.), valleys (similar to moorlands, but lacking distance cues and possibly including some specific image echoes) and forests (typified by increasing reverberation as the distance of the source from the listener increases). Such real, or apparently real, acoustic spaces may be re-created in the studio. For example, using the stereo projection of sounds on two loudspeakers we may separate sound sources along a left–right axis, creating a sense of spatial width. Simultaneously we may create a sense of spatial depth by using signals of smaller amplitude, with their high frequencies rolled off. In this way, depth is added to the image and we create an effective illusion of two-dimensional space. This illusion is enhanced if sound-objects are made to move through the virtual space. Depending on which type of acoustic space we wish to re-create, we might, for example, add more reverberation to these sources the more distant they appear to be. A more recent development, the digital technique known as 'convolution', allows us to impose in a very precise manner the acoustic characteristics of any pre-analysed sound environment upon a given sound-object.

Sound recording and the presentation of recorded material has, however, brought with it a number of other acoustic spaces which are conventions of a mode of presentation. These will be referred to as 'formalized' acoustic spaces to distinguish them from the real acoustic spaces we have previously been discussing. There is, of course, no clear dividing line between these two categories because, as broadcast sound becomes an ever more present part of our real sound environment, it becomes possible to question, for example, whether we hear the sounds of an orchestra, the sounds of a radio (playing orchestral music), or the sounds of a person walking in the street carrying a radio (playing orchestral music)! These questions are not trivial when we come to discuss electroacoustic music, such as Stockhausen's *Hymnen*, which attempts to integrate existing recorded music into its sonic architecture. And in fact some sounds which have been modified by the technological means of reproduction might almost be accepted into the category of real acoustic space: for example the sound of a voice heard over a telephone or the sound of a voice heard over a distant distorting channel as with the sounds of voices transmitted from space.

The formalization of acoustic space is found in all kinds of contemporary music production. In the studio recording of rock-music albums, various formal manipulations of the acoustic space are taken for granted. The most obvious is the rebalancing of instruments by means of differential amplification. A soft-spoken singing voice or a flute may become much louder than a whole brass section. Electrical amplification permits 'crooners' to sing at a very low level, and hence adopt the vocal characteristics of intimacy only available at low-level amplitudes, while still being heard against a large orchestra. At the other extreme, singers often use effects of reverberation, echo (often used in a self-conscious, clearly non-natural way) and phasing (an effect that is confined in the natural world almost exclusively to our perception of aeroplanes passing overhead!) to distance themselves from the listener or appear larger than life. (Such techniques have parallels in certain preliterate cultures.) However, as these techniques are usually applied on an all-or-nothing basis in a particular song, and are used frequently, they cease to have any meaning as features of an imagined acoustic space and become conventions of a formalized musical landscape. At the present stage in the development of these devices, they can only become useful tools for the elucidation and elaboration of landscape in our perception of sonic art if we abandon the all-or-nothing usages and investigate the effects of transformations between different representations of the voice with a sensitivity to the concept of aural space.

This formalization of acoustic space does not apply merely to music. It is an aspect of the format of radio and television broadcasts. A typical formalized landscape is the disc jockey's presentation of rock music. The record ends or draws to a close, the voice of the disc jockey is heard. This is often mixed with special snippets of music or sound-effects (like 'logos' in sound) and spoken or sung linkage material ("keep the music playing") until the next song begins. This controlling voice, floating on a sea of musical snippets, has now become a broadcasting convention and has no specific landscape implications.

Finally, we might consider the case of electroacoustic compositions such as Stockhausen's *Telemusik* or *Hymnen* which use elements of recognizable recorded music within a sonic architecture using a wider sound palette. In both pieces the relationship between the finite portions of recognizable recorded music and the stream of electronically generated sound materials into which they are absorbed suggests a view from outside or above the cultural substratum from which the musical extracts originate. We are here in a sort of 'cosmic' media space, a musical process on which we are 'carried' and in which the 'distancing' or detachment from the real (no recognizable acoustic space, no recognizable real-world referents, although some kinds of natural processes are suggested in *Hymnen*) predisposes us to perceive the pre-recorded extracts (themselves heavily distorted) in a distanced way, as part of a broader 'cosmic' perspective. Such a generalized use of this sense of detachment tends to become accepted as convention as time goes by and this aspect of Stockhausen's aural metaphor may not survive (it may be perceived merely in formal terms) as our familiarity with electronic sound sources increases.

II. The disposition of sound-objects within the space

Given that we have established a coherent aural image of a real acoustic space, we may then begin to position sound-objects within the space. Imagine for a moment that we have established the acoustic space of a forest (width represented by the spread across a pair of stereo speakers, depth represented by decreasing amplitude and high-frequency components and increasing reverberation); we may then position the sounds of various birds and animals within this space. These sound sources may be static, individual sound sources may move laterally or in and out of 'depth', or the entire group of sound sources may move through the acoustic space. All of these are at least perceivable as real landscapes. If we now choose a group of animals and birds which do not, or cannot, coexist in close proximity, the landscape would, ecologically speaking, be unrealistic, but for most listeners it would remain a real landscape.

Let us now begin to replace the animal and bird sounds by arbitrary sonic objects. We might accomplish this by a gradual process of substitution, or even by a gradual transformation of each component of the landscape. At some stage along this process we begin to perceive a different kind of landscape. The disposition of the objects remains realistic (in the sense that we retain the image of the acoustic space of a 'forest') yet the sound sources are not real in any sense of the word. Here we have the first example of an imaginary landscape of the type 'unreal objects / real space'.

If we now take the original sound-objects (the animal and bird sounds) and arbitrarily assign to each occurrence different amplitudes and degrees of reverberation or filtering, we achieve a second but quite different kind of imaginary landscape of the type 'real objects / unreal space'.

A more extreme example of the ecologically unacceptable environment (!) described earlier, brings us to a third type of landscape. Imagine, for example, that, by appropriate editing and mixing procedures, we are able to animate a duet between a howler monkey and a budgerigar or a whale and a wolf; we have a landscape in which the sound sources are real and the perceived space is real, yet the relationship of the sound-images is impossible. This bringing together of normally unrelated objects in the virtual space created by loudspeakers is closely parallel to surrealism, the technique of bringing together unrelated visual objects in the space defined by a painting. I therefore propose to call this type of imaginary landscape ('real sounds / real space') 'surrealist'.

The change in the apparent disposition of sonic objects in the acoustic space may alter the perspective of the listener. For example, where various sound-objects move in different directions, or a single sound spins around the listener's position (in a quadraphonic space), the listener may reasonably assume himself or herself to be at a point of rest. If, however, sound-objects are spaced at various points of the compass around a quadraphonic space, and the entire frame of reference made to rotate, we might suggest to the listener that he or she is spinning in the opposite direction whilst the frame of reference remains still. At a simpler level, differences in amplitude and also timbral qualities caused by the closeness or distance of the microphone to the recorded object alter not only the listener's perceived physical distance from the source but also the psychological or social distance. With vocal sounds, depending on the type of material used, closeness may imply intimacy or threatening dominance, distance a sense of 'eavesdropping' or of detachment, and at various stages in between a sense of interpersonal communication or more formalized social communication. A similar kind of psychological distancing which parallels the social distancing may be experienced even in the case of inanimate sound-sources. To hear sounds from our normal acoustic experience in the same perspective that close-miking provides we would usually need to be purposefully directing our aural attention to the sounds (by, for example, bringing the sounding object very close to our ear). Listening to sounds recorded in this way can produce the effect of perception through an aural 'magnifying glass' and is quite different from our experience of normal acoustic perspective. Changes in aural perspective in the course of an event or extended composition are of psychological import for the listener and should not be composed in a purely formalist manner.

III. The recognition of individual sound-objects

First of all we should note that in our normal working lives our experience of the environment is a multi-media one. In particular we rely very heavily on the visual medium to assist in our recognition of objects and events. This recognition may be direct, in the sense that we see the object or event which emits the sound, or indirect, in the sense that a physical location (for example a railway station or a particular type of terrain) or a social occasion (such as a concert performance) may enable us to identify a (perhaps indistinctly heard) sound source. Without visual cues, however, we may still rely on contextualizing aural cues to aid our recognition of a source. These cues may not only affect our recognition of an aural image, but also our interpretation of the events we hear. As a simple example, imagine a recording of a vocal performance accompanied by piano. Imagine that the vocal performer uses many types of vocal utterance not normally associated with the Western musical repertory, such as screaming, glossalalia or erotic articulation of the breath. The presence of the piano in this context will lead us to interpret these events as part of a musical performance, perhaps the realization of a pre-composed score. The utterance will lie within the formalized sphere of musical presentation. If, however, we were instead to hear a similar recording in which the piano were not present and no other clues were given about the social context in which the vocalizations occurred, we might not be able to decide whether we were listening to a 'performance' in the above sense or 'overhearing' a direct utterance of religious ecstasy or the ravings of an insane person.

Certain sounds retain their intrinsic recognizability under the most extreme forms of distortion. A most important sound of this type is the human voice, and particularly human language, although the particular formant structure of the human voice in itself is especially recognizable to human beings. This is partly due to the immediate significance of the human voice for the human listener, but also to the unique complexity of articulation involved. The ability to produce a rapid stream of timbrally disjunct entities is uncharacteristic of any other source, except perhaps bird mimics of human beings, such as parrots, and makes the recognizability of human utterances especially resistant to many types of sonic distortion. Given a repetitive, stridulant sound in isolation, it would probably be difficult for us to identify it as a cricket, electronic oscillator or mechanical vibration, although if these three sounds were heard in quick succession we would probably be able to differentiate them by comparison. On the other hand, if a sufficiently 'realistic' acoustic landscape were created, of the kind of habitat in which crickets live, it would probably be possible for an electronic imitation of a cricket to be inserted without our recognizing its true origins. The human voice, however, can be recognized even when its specific spectral characteristics have been utterly changed and it is projected through a noisy or independently articulated channel; it is also notoriously difficult to imitate electronically.

Transformation; intrinsic ambiguity of aural space

With an understanding of various properties of the aural landscape, we can begin to build compositional techniques based upon transformations of the landscape. Digital technology offers us tremendous power to manipulate the inner substance of sounds so that transformations between different recognizable archetypes can be effected with great subtlety. For example, the computer language CHANT[4] permits us to generate a transition from a bell-like sonority to a male voice by simultaneously manipulating the relative amplitude and width of the formants and the rate of occurrence of impulsions (which, when sufficiently fast, correspond to the perceived pitch of the voice).

It is interesting to compare these aural transformations with similar transformations which might be effected in visual space (for example in animated cartoons). It would seem plausible that there is some relationship between the two techniques, but we should be wary of drawing too close a parallel. Thus, although the change 'book-slam' to 'door-slam' may be quite easily conceived in visual space, the processes involved in the transformation of spatial form are quite different from those involved in the transformation of the acoustic object. The fact that book-shape and door-shape are similar, and at the same time book-slam sound and door-slam sound are similar, is a happy coincidence in this case. There is no intrinsic correlation between geometrical shape (by itself) and sonic properties. The distinction is even more evident when we consider the transformation from my own work *Red Bird*[5] in which the syllable 'lis' (of 'listen') changes into the sound of birdsong. What could be the visual equivalent of this transformation? Is it a mouth which changes into a bird or some representation of breath which becomes a bird? It soon becomes clear that seeking such a parallel is rather fruitless. Visual transformation and aural transformation take place in different dimensions. They also have quite different qualities. We normally have little difficulty in recognizing a visual object, even at a glance. Rapid transformations between clearly recognizable objects are therefore quite simple to achieve in visual animation; the whole process has a very 'concrete' feel, a certain definiteness. Aural images, however, almost always remain a little ambiguous. We can never be sure that we have made a correct recognition, especially where the transformation of sound-objects denies us the normal contextual cues. Transformation in aural space therefore tends to have a 'dreamlike' quality removed from the concrete, distanced and often humorous world of visual animation. At the same time, although transformations of abstract forms or between recognizable and abstract forms may be achieved in the visual sphere, these tend not to have the same affective qualities as time processes taking place in the domain of sound. Landscape composition, therefore, has a quite different feel from the sphere of visual animation.

The concept of landscape has now been used extensively by composers of electroacoustic music. In Ferrari's *Presque Rien no. 1*, as already mentioned, an existing landscape is simply recorded on tape and concentrated in intensity by editing the events of a whole day into the space of 20 minutes. The landscape itself remains entirely realistic, but our attention to it is concentrated by the condensation of materials and also by the fact of its being brought into the concert hall or living room. In Larry Wendt's *From Frogs* we hear a landscape consisting of the songs of many frogs but in this case the songs are slowed down and hence the rich content of the spectrum becomes more apparent to our ears. The effect is similar to that of close perspective (recording sounds with the microphone placed extremely close to the source); we have a sense of entering into the inner details of the sound world.

In Alvin Lucier's *I Am Sitting in a Room*, on the other hand, the acoustic characteristics of the physical space in which the sound-object is first presented are brought to the centre of our attention by the process of natural resonant feedback. Lucier's approach to acoustic space is literal and objective. He is concerned with drawing our attention to, and utilizing, the acoustic characteristics of existing spaces, rather than establishing a virtual space through the medium of the loudspeaker. His work also tends objectively to demonstrate or elucidate these characteristics rather than attempt to utilize them for some other expressive purpose. In Michael McNabb's *Dreamsong*, however, we are concerned with transformations of virtual acoustic space. The transformations are neither simply relatable to existing acoustic spaces, nor do they relate to any conceivable or visualizable events in the real world. As its title suggests, we find ourselves travelling in a dream landscape which has its own logic. The further elucidation of this 'dreamlike' virtual landscape is the basis of the piece *Red Bird* which I will discuss more fully below.

Music and myth

Having established that considerations of landscape enter into our perception of sonic art and that representational sound-images are potential working material, what special implications does this have for the sonic artist? In particular, what forms may we develop based on our sensitivity to sound-images. In *Dreamsong* the transmutation of aural landscape is suggestive of the scenarios of dreams, whilst the mediation between the human voice and the 'pure', 'abstract' electronic sounds in Stockhausen's *Gesang der Jünglinge* points towards a metaphorical interpretation.

Gesang takes the metaphorical opposition of voice and abstract material and embeds it in a complex musical structure. Although this continues to contribute to our perception of the work, it is not further elaborated as *metaphor*. What would happen if we were to establish a whole system of relationships between sound-images, each having strong metaphorical implications? By articulating the relationships between the sound-images we could develop not only sonic structures (as in McNabb) but also a whole area of metaphorical discourse.

In 1973, having worked for some time on electroacoustic compositions which utilized sound-images within an otherwise mainstream conception of sonic structure, I decided to attempt to set up a sonic architecture based on the relationship between the sound-images themselves which would however remain compatible with my feelings about musical structure.

What I discovered through working on the piece *Red Bird* was that the twin concepts of transformation and gesture (which I have elsewhere[6] discussed in relation to non-representational sound-objects), may also be applied to the sound-image. On the one hand, sound-images of the voice, or animal and bird cries, have an intrinsic gestural content. More distanced sound materials, such as textures developed out of vocal syllables, may be gesturally articulated by appropriate studio techniques. Transformation now becomes the gradual changing of one sound-image into another with its associated metaphorical implications, and a landscape can be seen as a particular kind of timbre-field applying to the space of sound-images. These parallels are not, of course, precise, but they do form the basis of a meeting ground between musical thinking and a discourse using sound-images as concrete metaphor.

In his book *The Raw and the Cooked*[7], Lévi-Strauss draws certain interesting parallels between the structure of music and the structure of myth. At one level, in adopting a structural approach to the analysis of myth, Lévi-Strauss calls upon structural categories from the tradition of Western music (such as 'sonata form' and 'double inverted counterpoint'). However, he also implies a deeper relationship between music and myth. Using people, objects and wild animals with a particular significance for the group, the myth illuminates more abstract relationships and categories of thought. At the same time the myth gains its power from its unfolding in time. The way the myth is told is of great importance. The parallel with conventional musical structures is obvious and in fact Lévi-Strauss points to Wagner as the first person to attempt a structural analysis of myth. The fact that Wagner used music as his medium is, for Lévi-Strauss, no coincidence.

Wagner's methodology establishes a relationship between delimited musical structures (leitmotifs) and people, objects or ideas, primarily through association (at some stage musical object and its referent are juxtaposed). By developing these musical objects and interrelating them, he is able to carry on a discourse which is not subject to the spatial and temporal limitations of the opera stage. This discourse is to do partly with unspoken 'emotions' and partly with metaphor.

Using sound-images in the virtual space of the loudspeakers, we can create a world somewhere in between the concreteness of the opera staging and the world of musical relationships. We do not need to associate a musical object with, for example, a bird and thence with a metaphorical meaning; we may use the sound of a bird directly. And the concreteness of theatrical staging is replaced by a dreamlike landscape hovering between musical articulation and 'real-world' events. Having drawn these parallels, however, from this point on there will be little in common between our perception and conception of sound-image composition and that of Wagnerian opera!

In looking at Stockhausen's *Gesang* we have already noted that sound-images may be used metaphorically, although in this particular case this metaphorical use may not strike us immediately, as other features of the sonic architecture are more strongly articulated and apparent to our perceptions. It is interesting to note, however, that even in this case the metaphorical interpretation depends on the existence of a transformation, the mediation between the sound of a voice and the electronic sound. In a similar way, we might consider using the aural image 'bird' (as in *Red Bird*) as a metaphor of flight (and hence, perhaps freedom or imagination), although in itself the sound of a bird need conjure up no such metaphorical association. If, however, we now make the sonic transformation from 'lisss' to birdsong (in *Red Bird* the syllable 'lisss' is understood to be from the phrase 'listen to reason'), the voice 'takes flight' so to speak. The metaphorical link with the concept 'imagination' is suggested. If this transformation is set within a whole matrix of related and transforming images the metaphorical implications become increasingly refined and ramified. Similarly, the sound of a mechanically repetitive machine has no implicit metaphorical implications, but a 'word-machine' made out of syllables of the phrase 'listen to reason', and the relationship between the normally spoken sentence and the word-machine, begin to establish a metaphorical dimension in our perception of the sound-image. We may gradually establish a network of such relationships, defining a rich metaphorical field of discourse, and, just as a change of contextual cues may alter our interpretation of a sound-image, so it may also have metaphorical import. In *Red Bird*, the sound-image 'bellows/water-pump' may be interpreted as the functioning of a machine or the functioning of a human body, and when our perception of it changes from one to the other a metaphor is implied. (These are relatively crude attempts to describe the aural landscape which is always much more ambiguous than the visual.) The listener may of course deny or blank out the metaphorical implications, but this is possible with all other art forms which use metaphor.

In putting together a sonic architecture which uses sound-images as metaphors, we are faced with a dual problem. We must use sound transformations and formal structures with both sonic impact and metaphorical import; we must be both sonically and metaphorically articulate. Using concrete metaphors (rather than text) we are not 'telling a story' in the usual sense, but unfolding structures and relationships in time; ideally we should not think of the two aspects of the sound landscape – the sonic and the metaphorical – as different, but as complementary aspects of the unfolding structure. This fusion of conception took place slowly and unnoticed in my mind during the composition of *Red Bird*.

In order to build up a complex metaphoric network we need to establish a set of metaphoric primitives which the listener might reasonably be expected to recognize and relate to, just as in the structure of a myth we need to use symbols which are reasonably unambiguous to a large number of people. Such criteria influenced the choice of metaphoric primitives for *Red Bird* but the use of intrinsically esoteric referents *à la* T. S. Eliot would not have been appropriate in this context. The four basic sound types used in *Red Bird* are Words (especially 'listen to reason'), Birds, Animal/Body and Machines. While in certain cases these categories are quite clearly distinguishable, ambiguities do arise and are used intentionally. For example, non-linguistic vocal utterances (from a human or animal voice) may approach linguistic forms and vice versa; a repeated high-frequency glissando may be taken to be of ornithological or mechanical origin; articulated mid-ranged pitched material may be both bird-like and animal-like. Each symbolic type is chosen because it either has a conventional symbolic interpretation (birds: flight, freedom, imagination; machine: factory, industrial society, mechanism) or such an interpretation can be easily established. (The phrase 'listen to reason' points to itself.)

The situation is already, however, more complicated than this might suggest. For example, the phrase 'listen to reason' is open to fragmentation and permutation and many different kinds of gestural articulation, which means that its semantic surface may be utterly transformed or even negated. More importantly, there are two particular kinds of landscape in which the sound-images of the piece may be placed. These may be described as the 'garden' landscape and the 'reason' landscape. Apart from the sound-images involved, the former is characterized by a sense of the coexistence of the sound-images in acoustic space and time (individual images being subordinated to a rigid, rhythmic structure). In fact, the machines which inhabit this landscape are made up either from phonemes or body-like visceral sounds, whilst the squeaks and squeals of the machinery's operation are vocal, animal or bird noises.

The sound-images used in a landscape may be organized to suggest different interpretations of the landscape, or even different interpretations of the sound-images themselves. The garden landscape organizes the sound-images in a naturalistic way. Although the juxtaposition of species is ecologically impossible, our perception is of a 'natural', if somewhat dreamlike, environment. Conversely, in the 'Bird-Cadenza' of *Red Bird* the material is organized according to formal criteria derived from musical experience, but because of the sound-imagery used (various kinds of birdsong), the percept of a natural environment has not been destroyed.

In the case of the word-machine our attention is not focussed upon the phonemic constituents of the sound as language; our interpretation if of a mechanical, not a verbal, sound-image. We hear phonemes *as* a machine. In other cases phonemes are subjected to a special kind of organization, such as a dense and diverging texture suggestive of the flocking of birds; we hear phonemes *as if* they were flocking birds. In other cases, the phoneme 'rea' of 'reason' *gradually becomes* the sound of an aggressively barking dog, or the word 'reasonable' *immediately becomes* the sound of bubbling water; the final phoneme 'ble' apparently bursts to reveal bubbling water.

Is there a natural morphology of sounds?

While investigating sound-objects from the point of view of the landscapes they create, it is interesting to reconsider certain categories from the acousmatic description of sound-objects given by the Groupe de Recherches Musicales. In particular the category of 'continuation' refers, as its name suggests, to the way in which a sound-object may be continued in time. Three basic categories emerge: the Discrete, the Iterative and the Continuous. Discrete continuation describes such sounds as a single (unresonant) drum-stroke or a dry pizzicato on a string instrument. Iterative continuation applies to a single-note 'trill' on a xylophone (the sustaining by rapid attack of a sound which would otherwise be discrete), a drum-roll, the stream of rapid clicks produced by vocal 'grating' (also known as 'vocal fry'), or a bowed note on a double bass string which has been considerably slackened off. Continuous continuation applies to a sustained note on a flute, a synthesizer or a bell.

These three types of continuation imply something about the energy input to the sounding material and also about the nature of the sounding system itself. We may, in fact, say about any sound-event that it has an *intrinsic* and an *imposed* morphology. Most sound-objects which we encounter in conventional music have a stable intrinsic morphology. Once the sound is initiated it settles extremely rapidly on a fixed pitch, a fixed noise-band or more generally on a fixed 'mass' as in the case of bell-like or drum-like sounds with inharmonic partials. Furthermore, most physical systems will require a continual (either continuous or iterative) energy input to continue to produce the sound. Others (such as bells or metal rods), however, have internal resonating properties which cause the sound energy to be emitted slowly with ever-decreasing amplitude after an initial brief energy input. We may therefore, from the point of view of landscape, split the category of continuous continuation to give, on the one hand, sounds such as those of flute or violin where there is a continuous input of energy (continuation through imposed morphology) and, on the other hand, sounds where the continuation is due to the physical properties of the sounding medium (continuation through intrinsic morphology).

The reason for making this distinction is simply that the imposed morphology tells us something about the energy input to the system (and ultimately relates to what I have called the 'gestural structure of sounds'). Clearly we can gain most information about this energy input where it is continuous and least where it is in the form of an initiating impulse. Where energy (mechanical, hydraulic, aerodynamic or electrical) is continuously applied to the system, we can follow its subtle fluctuations. The sounding system is gesturally responsive. However, where a sound-event is initiated by an impulse (drum-stroke, bell) very little gestural information can be conveyed (effectively, only a difference in loudness relating to the force of the impulse). Iterative continuation is ambiguous in this respect. Iteration may be entirely an aspect of the applied force (as in the case of the xylophone 'trill'), purely an aspect of the physical nature of the medium (vocal fry or slack double bass strings), or an interacting mixture of the two (a drum-roll).

Clearly, on an analogue electronic synthesizer, for example, we can generate events of any kind without actually supplying any immediate energy input from our own bodies, but the mode of continuation (attack-structure and articulation etc.) of a sound will tend to be read in terms of the physical categories I have described. The distinction between continuous and impulse-based excitation, for example, is not a mere technical distinction but relates to our entire acoustic experience and 'tells us something' about the sound-object even though it may have been generated by an electrical procedure set up in an entirely cerebral manner. We can, of course, transcend these categories of the physical experience of sound-events, but I would suggest that we do so in the knowledge that this background exists. In a similar way we may generate glossalalia through, for example, the statistical analysis of letter frequencies in texts, but the reader will always hear or read the results against the background of a knowledge of one or several languages. The forms of sound-objects are not arbitrary and cannot be arbitrarily interrelated.

Composers who have weighted their activities towards live electronics rather than studio-based synthesis seem to me to have been strongly affected by the fact that a morphology imposed upon electronic sound-objects through the monitoring of performance gesture can be much more refined and subtle than that resulting from intellectual decisions made in the studio. The directness of physiological-intellectual gestural behaviour carries with it 'unspoken' knowledge of morphological subtlety which a more distanced intellectual approach may not be aware of. This is not to say that theorizing cannot lead to interesting results, but implies that it can lead to a loss of contact with the realities of the acoustic landscape.

Even where the imposed morphology is a mere impulse, the loudness of the sound carries information about the force of that impulse. The association of loudness with power is not a mere cultural convention but loud sounds have often been used to project political potency (massed orchestras and choruses, the musicians of the Turkish army etc.). As far as we know, overall continuous changing in dynamic level (crescendos, diminuendos) were an invention of the Mannheim school of symphonic composition in the eighteenth century. The formalistic assignment of a series of different dynamic levels to musical objects, which was experimented with in the total serial aesthetic, leaves a sense of arbitrariness or agitation (neither of which is usually intended) because it ignores the landscape basis of our perception of loudness.

Sounds undergoing continuous excitation can carry a great deal of information about the exciting source (this is why sounds generated by continuous physiological human action – such as bowing or blowing – are more 'lively' than sounds emanating, unmediated, from electrical circuits in synthesizers). The two natural environmental sounds which indicate continuous excitation, the sound of the sea and that of the wind, have an interesting temporal morphology (which may relate to the symbolic association of these sounds). In the case of the sea, the excitation (the pull of the moon's gravity) may be regular, but the form of the object (the varying depth of the sea bottom) results in a somewhat unstably evolving (intrinsic) morphology. The sound 'of the wind' is usually in fact the sound of something else animated by the motion of the wind and in this case it is the exciting force itself (the wind) which varies unpredictably in energy, which gives the sound its interestingly evolving (imposed) morphology. As Murray Schafer has pointed out in his book *The Tuning of the World*[8], "It is only in our present technological society that continuous sounds which are completely stable (the hum of electrical generators etc.) have come to be commonplace." The ability of the synthesizer to generate a continual stream of sounds says something about our society's ability to control energy sources, but if we take this continuous streaming for granted, it, like the humming of machinery, tends to lose any impact it might have had on the listener. The machine has no intentions and therefore it inputs no gestures to its sounds. The synthesizer can sound the same way!

One important critique of the acousmatic analysis of sound-objects is that it reduces the two dimensions of imposed (gestural) morphology and intrinsic morphology to a single dimension, even though the distinction between these two is not totally clear-cut and in the virtual acoustic space of loudspeakers the problem of sound origins can be perplexing. I would argue that the two dimensions continue to enter into our *perception* of sound-objects. Different kinds of intrinsic morphology affect us differently and this is something to do with the assumed physicality of the source (which is not the same thing as source-recognition). Imposed morphology we react to more directly, having an immediate relation to the workings of our own physiological-intellectual processes.

Most musical instruments have a stable intrinsic morphology. When energy is input in a steady stream or as an impulse, they produce a sound-object of the attack–resonance type. There is an initial excitation which generates a distinct spectrum (either pitched, inharmonic, or noise-based) which then dies away in amplitude either rapidly or with varying degrees of sustainment. Not all physical objects, however, behave in a similar fashion. If we apply a stream of air whose pressure gradually increases, in the case of a flute we find that we set up a given fixed pitch which becomes louder (until we reach a 'catastrophe' point at which the note changes to a different harmonic of the fundamental), whereas in the case of a siren the pitch slides upwards as the pressure is increased. The human voice is an interesting case in this respect because, although in music it is usually used to model the musical instruments we use, in moments of extreme stress it tends to react more like the siren (the scream).

If, however, we take sound-objects whose intrinsic morphology is very complex or unstable, how can we relate to these? Are they merely formless or random? I would propose that there are a number of archetypes which allow us to classify these complex sounds perceptually, such as Turbulence, Wave-break, Open/Close, Siren/Wind, Creak/Crack, Unstable/Settling, Shatter, Explosion, Bubble. And although there is not space in this article to enlarge on these ideas I would suggest that it may even be possible to extend this kind of analysis to phenomena where many individual sound sources are amassed, for example the Alarum (when a colony of animals or birds is disturbed the resulting mass of individual sounds has a very characteristic morphology), or Streaming effects (certain changes occurring is continuous streams of sounds may perhaps be related to models developed in catastrophe theory).

In conclusion it seems clear that the various aspects of sound landscape we have discussed offer a broad and fascinating area for musical exploration and development. The sophisticated control of this dimension of our sonic experience has only become possible with the development of sound-recording and synthesis and the control of virtual acoustic space via sound projection from loudspeakers. It would certainly be foolish to dismiss this new world of possibilities on some *a priori* assumption that they are not of 'musical' concern. In fact, any definition of musical activity which does not take them into account must, from here on, be regarded as inadequate.

4

Spectro-morphology and Structuring Processes[1]

Denis Smalley

Introduction

The development of Western music in the twentieth century is dominated by an historic bifurcation in musical language: tonality with its metrically organized harmonic and melodic relationships has continued to be the vernacular language, absorbed unconsciously from birth, while the other fork, in its most recent guise, is represented by spectro-morphology[2]. Spectro-morphology is an approach to sound materials and musical structures which concentrates on the spectrum of available pitches and their shaping in time. In embracing the total framework of pitch and time it implies that the vernacular language is confined to a small area of the musical universe. Developments such as atonality, total serialism, the expansion of percussion instruments, and the advent of electroacoustic media, all contribute to the recognition of the inherent musicality in all sounds. But it is sound recording, electronic technology, and most recently the computer, which have opened up a musical exploration not previously possible. Spectro-morphology is a way of perceiving and conceiving these new values resulting from a chain of influences which has accelerated since the turn of the century. As such it is an heir to Western musical tradition which at the same time changes musical criteria and demands new perceptions.

Spectro-morphology finds its true home in electroacoustic music but it is not imprisoned there. Although instruments may also be moulded in a spectro-morphological manner, traditional wind and string instruments, harmonic in their spectral makeup, were conceived and developed for an harmonic music. Even if modern performing techniques seem to have enabled us to escape from harmonic confines, it is a temporary and illusory freedom subverted by the traditional nature of instruments. The future of live performance must lie with new instruments.

The voice, because of its intimate human links, can never fade from musical usage. Through language it has been infused into over a thousand years of Western music including purely instrumental music.[3] It is therefore inevitable that the voice and language continue to lie at the heart of vernacular music, while along the other fork their influence has weakened as spectro-morphology has seeped into modern musical thinking. As sound sources, however, language and vocal sounds have discovered a new significance as a result of their contact with electroacoustic media.

Even if the actual sound of structures based on a spectro-morphological approach often appears to leave voices and instruments far behind, their formative influence nevertheless persists through gesture: the spectral shapes and shape-sequences created by the energy of physical and vocal articulation. Though the internal spectral behaviour of sounds may no longer mirror overtly the inspiration of instruments and voices, tangible links with humanity demand to be preserved through gesture.

Recently the sounds of environmental phenomena have been accorded equal status with voice and instrument. The acceptance of environmental sounds completes the sounding models for musical composition: the sounds of language and human utterance in general, natural phenomena, sounds intentionally created by human agency (not only those of musical instruments), and unintended sounds of human origin. The musical exploration of these sources and their imaginative extensions through signal processing and synthesis are the business of the electroacoustic arts.

* * *

Listeners can only apprehend music if they discover a perceptual affinity with its materials and structure. Such an affinity depends on the partnership between composer and listener mediated by aural perception. Today we continually need to reassert the primacy of aural experience in music. The heritage of twentieth-century formalism and the continuing propensity of composers to seek support in non-musical models have produced the undesirable side-effect of stressing concept at the expense of percept. Borrowing concepts from non-musical disciplines is common and can be helpful, but unless concept is cross-checked or mitigated by the ear it is always possible that the listener will be ostracized. Aural perception is fragile, fickle, empirical, and thus presents a threat to those musicians and researchers who have difficulty in coming to terms with the insecurity of their subjectivity. The primacy of perception is unassailable since without it musical experience does not exist.

We need to concentrate our attention on musical perception and in particular the now extensive pool of electroacoustic music at our disposal. It is my experience through teaching, attending and programming concerts, and talking to electroacoustic music performers and composers that there is a remarkable consensus about the efficacy or failure of particular musical works, thus indicating an instinctive evaluation of the newer spectro-morphological values. The practice of listening, and the perceptive observation of the listening process must therefore form the foundation of any musical investigation which seeks to explain the workings of spectro-morphology.

* * *

The lack of shared terminology is a serious problem confronting electroacoustic music because a description of sound materials and their relationships is a prerequisite for evaluative discussion. In searching for appropriate words we are obliged to borrow non-musical terms because the circumscribed vocabulary invented for purely musical explication is too limited for spectro-morphological purposes. Such semantic borrowings immediately indicate that music involves mimesis: musical materials and structures find resemblances and echoes in the non-musical world. These links with human experience may be obvious, tangible and conscious, or covert, elusive and unconscious. However, to find out what happens in the life of a sound or sound structure, or what attracts us about a sound quality or shape, we must temporarily ignore how the sound was made or what caused it, and concentrate on charting its spectro-morphological progress. For example, in trying to describe the sound of an approaching car we should have to forget that it is a car and ignore everything associated with 'car-ness', confining our aural observations to discovering how the spectrum of the sound changes in time. This investigative process is known as *reduced listening*[4] – 'reduced' because purposely rejecting the sound's source reduces the scope of normal musical experience. Such an abstract listening attitude should not be regarded as a purely investigative device divorced from real musical experience. One can imagine a context where the spectral design of the car's sound could be associated with other sounds of related shape. In this way the composer could manipulate the context so that the listener is drawn into following the abstract aspects of the sound's design rather than thinking about the significance of cars as objects. On the other hand a musical context could be created where the car's sound is used to make a statement about cars as cultural symbols. Ultimately, in explaining the role of the car's sound in its musical context, we should want to explore both aspects of its meaning.

All sounds possess this dual potential – the *abstract* and *concrete* aspects of sound[5] – and all musical structures are balanced somewhere between the two, although exactly how they are found to be balanced can vary greatly among listeners. This is because all listeners have considerable practice at the concrete aspect in daily life, while an abstract approach needs to be acquired. However, the listener used to a more abstract perceptual attitude can easily disregard the mimetic dimension of interpreting sounds. Balancing abstract and concrete attitudes is therefore a question of both competence and intention.

This balance is further complicated because the abstract and concrete aspects are not always what they seem. Music is always related in some way to human experience, which means that mimesis is always at work even in music regarded as abstract, though such mimesis is notoriously difficult to explain, particularly as language often proves an inadequate filter for interpreting musical experience. On the other hand, a musical context which appears to depend entirely on mimetic impact is equally deceptive. The power of a concrete sound-image to portray things, events or psychological circumstances, rests not just on the immediacy of the images themselves but on how the sounds are constructed and combined – their spectro-morphology – and that involves using reduced listening to investigate the more abstract dimension.

The terminology in this chapter often evokes extra-musical analogies and many words were selected because of their associations. In adopting a spectro-morphological approach we should use reduced listening as the main investigative strategy, remembering that analogical terminology simultaneously invites necessary mimetic interpretations. We shall first examine how pitches are combined to form different sound types: spectral typology.

We can then follow the ways in which spectral types are formed into basic temporal shapes or morphologies. This leads us to a broader discussion based on the idea of motion: the directional tendencies of sound shapes and combinations of shapes. We can then consider structuring principles based on how the listener experiences musical time and motion.

Spectral typology

The term 'spectrum' encompasses the totality of perceptible frequencies. It replaces the former division of the frequency domain into 'pitch' and 'timbre' – terms too closely associated with traditional notions about their functions within harmonic music. The idea that 'notes', the bearers of pitch information, are clothed in timbral hues is not eliminated; it is merely located in the wider perspective of spectral types.

Spectral typology cannot realistically be separated from time: spectra are perceived through time, and time is perceived as spectral motion. Therefore only a limited, preliminary discussion is possible until morphology and motion are approached.

* * *

Three spectral types provide central reference points for spectral identification: *note*, *node*, and *noise* (see Fig.1). The note type, which is concerned with the perception of a discrete pitch or pitches, can be subdivided into three categories: *note proper*, *harmonic spectra* and *inharmonic spectra.*

The *note proper* embraces traditional pitch perception: absolute pitches, intervallic and chordal combinations. This means that even though the note

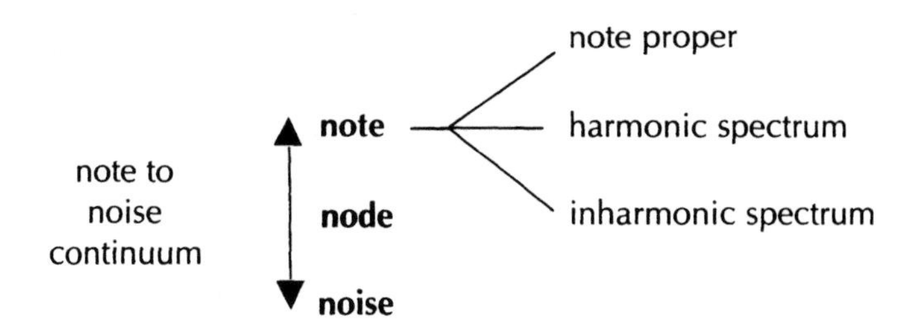

Figure 1: Spectral typology

may be spectrally coloured we are more interested in its fundamental pitch than its overtones. In the tonal system notes are the prime information carriers. If they are used to any extent, particularly in intervallic or harmonic contexts, the tonal system may be evoked and it becomes more difficult to coax the ear into observing concurrent spectral qualities, such is the overriding primacy of note perception both as a natural and cultural phenomenon. However, we now realize that the borders between pitch and 'timbre' are very imprecise, and subject to drifting perceptual demarcations among which we may fluently commute if the musical context permits, and if we have acquired appropriate discriminatory skills. Tonal attitudes are therefore not abandoned or segregated but subsumed, and we find a place for the note proper in territory at one extreme of the typological continuum.

The step from note to *harmonic spectrum* is not decisive. In an harmonic spectrum intervals of the harmonic series govern. Whether the fundamental tone or the harmonic relationships of the spectrum attract perceptual attention depends on discriminatory habits and musical context. Context is crucial for spectral interpretation. In a traditional instrumental or vocal note the harmonic spectrum is balanced and fused to such an extent that the spectral components are largely indiscernible to, or at least ignored by, the ear. If fusion loosens its grip, harmonic focus becomes possible. Once an harmonic spectrum is perceptible above a fundamental its spectral components can be featured as compositional values. This focal transition can be demonstrated by listening to a low piano note. Heard at a normal listening distance the note proper predominates. If amplified or listened to more intimately the note's spectral components and their temporal shaping become apparent. The focal change from the note proper to the internal behaviour of components has furnished a new musical potential, since the electroacoustic medium makes viable the composition, decomposition, and development of spectral interiors.

Such a statement is even more significant for *inharmonic spectra* since without technology's assistance inharmonic spectra could never have become truly viable musical materials. Inharmonic spectra are modelled on the behaviour of, for example, many metallic sounds whose spectral components are generally unrelated to the harmonic series. Inharmonic spectra are not distributed in the predictable systematic order demonstrated by harmonic spectra. Their dispersed components often resist fusion and may therefore be perceived from a variety of angles provoking a fruitful ambiguity of focus. An inharmonic spectrum may contain intervals which evoke tonal references;

simple inharmonic spectra may be interpreted as close relations of harmonic spectra; more complex examples may include harmonic and inharmonic intervals alongside nodal densities which defy definition. The composer can invest significance in a particular focal angle by manipulating component behaviour towards one of a variety of spectral outcomes: inharmonicities, harmonicity, tonal intervals, or the more compact densities of nodal and noise typologies. Inharmonic spectra are ambiguously multi-dimensional.

A *nodal spectrum* is a band or knot of sound which resists pitch identification. Certain percussion instruments provide familiar models. A cymbal heard at a distance is perceived nodally, in that we tend not to identify a definite pitch but perceive its qualities as a unified, metallic, rich resonance. If we listen more closely, however, or amplify the cymbal, we may discover an internal spectral combination of note, harmonic, inharmonic or nodal components. A nodal spectrum can also be regarded as a sound density whose unified compactness makes it difficult to hear its internal pitch structure. A note-cluster furnishes a simple example.

The nodal spectrum lies towards the noise boundary of the *note–noise continuum*. The density of a *noise spectrum* is so compressed that it is impossible to hear any internal pitch structure. If we perceive internal qualities they are likely to be granular or particled motions, not pitched phenomena. Noise is not a monochrome but a variegated phenomenon, no less diverse than the other spectral types. Natural models are provided by wind and sea.

We pass through the blurred buffer zone between note and noise as a result of increased spectral density and compression. This may be an inherent property of the chosen sound materials or it may be the outcome of composed spectral superpositions. We shall highlight this buffer zone by calling it the *pitch–effluvium continuum*. Effluvium refers to the state where the ear can no longer resolve spectra into component pitches. Confronted by an effluvial state the listener needs to change focal strategy as aural interest is forced away from charting the pitch behaviour of internal components to follow the momentum of external shaping. Thus context changes the level at which the ear can respond to the musical structure. We shall return to this theme in discussing structure.

The structural potential of spectral attributes beyond the note proper is harnessed by comparing, relating, and transforming spectral types and their combinations. The note proper belongs to the unique cardinal system based on the ability of the ear to perceive absolute pitch values. If there is any spectral system to be discovered or created beyond the note proper it is an ordinal one, based only on relational degrees, not on absolutes.[6]

To realize the full musical potential of newer spectral attitudes we must now consider their temporal shaping, known as *morphology*.

Morphology

The morphologies of the instrumental note are a convenient introduction to concepts of temporal shaping, firstly because they are familiar to everyone and more significantly because they are sounding extensions of human action; there is a causal relationship between the action of breath or physical gesture and the consequential spectral and dynamic profiles. The effect of long-term conditioning has resulted in a reference-pool of aurally acceptable sound-objects which continue to exert both conscious and unconscious influence on electroacoustic composition.

During execution of a note, energy input is translated into changes in spectral richness or complexity. When listening to the note we reverse this cause and effect by deducing energetic phenomena from the changes in spectral richness. The dynamic profile articulates spectral change: spectral content responds to dynamic forces, or conversely, dynamic forces are deduced from spectral change. This aural congruence of spectral and dynamic profiles, and their association with energetic phenomena, are the substance of everyday perceptual practice. In the environment, when a sound approaches the listener its spectral and dynamic intensity increase at a rate proportional to perceived velocity. Moreover, the increase in spectral intensity permits the revelation of internal spectral detail as a function of spatial proximity. Musical sounds are inextricably bound up with this experiencing of time passing as interpreted through changes in spectral space, even when the sounds are not actually moving through space. If these natural fundamentals of sound perception are ignored in the composing of morphologies, in the structuring process, and in spatial articulation of structures, the listener can instinctively detect a musical deficiency. The evolution of spectral and dynamic change therefore works within tolerances naturally determined by aural experience. Working imaginatively with these tolerances lies at the heart of electroacoustic compositional skills and judgment, and a failure to appreciate their crucial importance frequently accounts for the poor response to electroacoustic works.

*   *   *

We may discern three *morphological archetypes* at the source of instrumental sounds: the *attack-impulse*, the *attack-decay*, and the *graduated continuant* (see Fig. 2). Their profiles can be notated using symbols. Each symbol outlines three linked temporal phases: *onset*, *continuant*, and *termination*. The symbol's vertical dimension represents spectral richness and dynamic level which are considered congruent.

The *attack* or *impulse* archetype is modelled on the single detached note – a sudden onset which is immediately terminated. In this case the attack-onset is also the termination.

The *attack with decay* is modelled on sounds whose attack-onset is extended by a resonance (a plucked string or a bell, for example) which quickly or gradually decays towards termination. The closed symbol represents the quicker decay which is strongly attack-determined. The open attack symbol with separated termination symbol reflects a more gradual decay where the ear is drawn away from the formative influence of the attack into the continuing behaviour of the sound on its way to termination. In the latter version we become aware of an equilibrium among the three linked phases, whereas the attack-impulse archetype is entirely weighted towards the onset.

The third archetype is the *graduated continuant* which is modelled on sustained sounds. The onset is graduated, settling into a continuant phase which eventually closes in a graduated termination. The onset is perceived as a much less formative influence than in the other two archetypes. Attention is drawn to the way in which the sound is maintained rather than to its initiation.

From these central archetypes a wide and subtle variety of temporal articulations is generated. For example, there is a great variety of onsets between a clear-cut attack and a tentative graduated onset, and we are aware

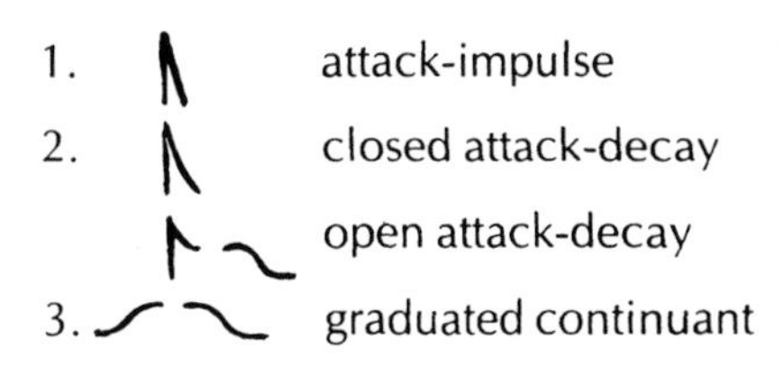

Figure 2: Morphological archetypes

of the intricacy of onset characteristics from instrumental performance practice. Indeed such characteristics serve to identify the personality of individual interpreters and the discriminating ear has proved sensitive to detailed spectral changes.

Departing from traditional reference points we may extend the archetypes into a broader listing of *morphological models* (see Fig. 3). Firstly it is useful to add a second graduated continuant type whose onset and termination phases are more rapid than those of the archetype: the *swelled graduated continuant*. Secondly, we include linear onsets and decays. Perfect linearity is normally less acceptable to the ear as it can sound too mechanical or artificial; absolute linearity is therefore more often synthetic than natural. Finally, we introduce the reversed versions of onset phases (which should not be confused with the literal reversal of sounds in tape composition). Terminations are not always silent closes. A change of spectral and dynamic direction during the continuant phase can initiate an increase in tension towards the termination. The context leads us to believe that we may be heading into a new onset. Thus a termination could also act as an onset.

Such a change of direction during the continuant phase of the open morphologies introduces the concept of *correspondence*, a point or stage in

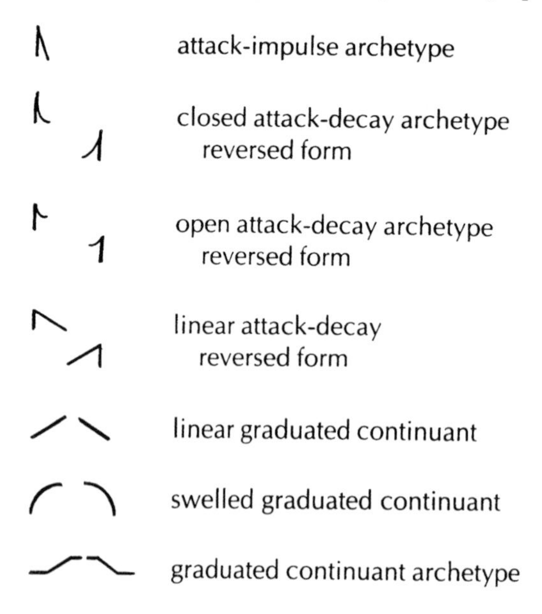

Figure 3: Morphological models

time where a morphological shift takes place – a kind of morphological modulation. Let us elaborate a hypothetical example of this process by taking the second archetype and expanding it into a structural unit. Attention is immediately drawn to the attack-onset, the impact of which sets in motion a resonating spectrum which we might expect to decay gradually. The further away we travel from the attack point the more we may be coaxed into observing how the spectral components of the continuant phase proceed, and the less interested we shall be in the genetic role of the initial impact. At this stage the composer may intervene in what so far has been a natural process by developing and changing the course of the spectral components. Through such intervention the expectation of decay can be postponed, diverted, or avoided, and the structure's orientation altered. The consequent spectral expansion cannot continue indefinitely, and a terminal strategy will eventually be invoked either to bring the structure to a close or to merge it into another structure.

But morphologies are not just isolated objects. They may be linked or merged in strings to create hybrids. Figure 4 enables us to show further possible correspondences during a *morphological string*. The open continuant phases permit correspondences through merging, as we found in the example based on the second archetype. Merged correspondences may also occur through the cross-fading of termination and onset, or more rapidly as a consequence of a reversed onset-termination. Of course morphological change may also occur through interpolation.

The stringing of attack-impulses creates perceptual circumstances with important structural consequences. By compressing the distance between

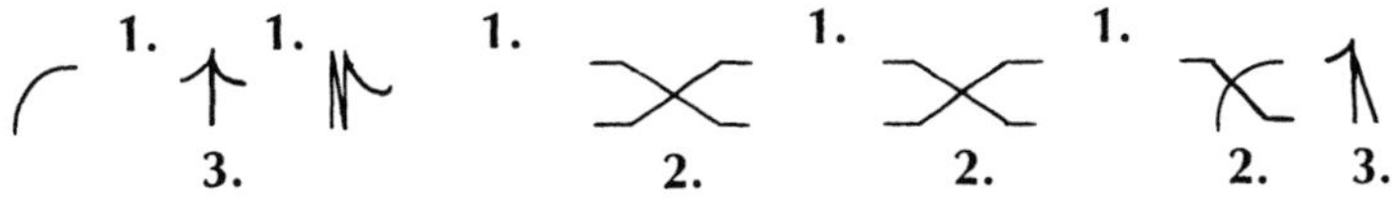

1. open continuant phases

2. merged correspondences through cross-fading

3. reversed onset-terminations leading to new onsets

Figure 4: Morphological stringing

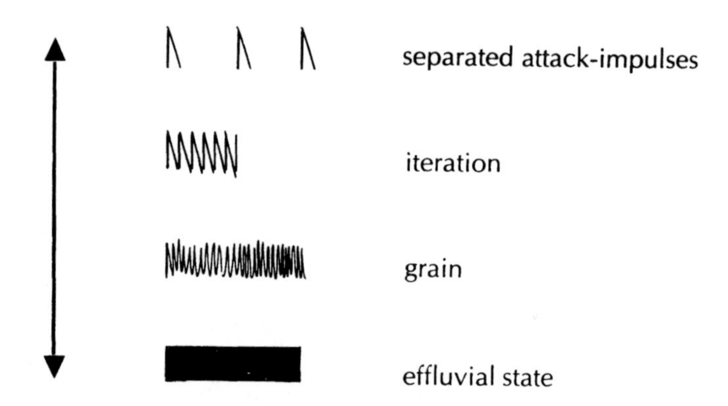

Figure 5: The attack–effluvium continuum

attack-impulses we enter the *attack–effluvium continuum*, a temporal counterpart of the *pitch–effluvium continuum* which we met under spectral typology (see Fig. 5). At the attack end of the continuum the separation between repeated attack-impulse objects is preserved. The first stage of compression creates *iteration*, where linked attack-impulses are perceived as a unified object. As the attack-impulses become more compressed we pass from *iteration* to the perception of *grain* where the once individual impulses have lost any vestiges of separate identity. A final compressive act will squeeze out any granular characteristics. Moving through the continuum the ear eventually fails to resolve the once separate components, and must therefore turn its attention to the morphologies which shape the broader stretches of structural motion. Thus, as with the pitch–effluvium continuum, contextual circumstances conspire to change the level at which the ear can respond to the musical structure. The musical consequences of pitch–effluvium and attack–effluvium are therefore the same, although the means of approaching the effluvial state are different.

* * *

It has already been pointed out in elaborating the second archetype into a large structure that the morphological archetypes and models are not confined to note-like dimensions. We may consider the tension design of the onset–continuant–termination relationship as a reflection of larger-scale structural focusses and functions, and vice versa. Thus the internal stresses of the morphological archetypes are projected into higher structural levels: the spectral and dynamic tensions inherent in the sounding extensions of gesture are the foundations of musical structuring. We shall return to examine this more closely once spectro-morphological observations are complete.

Motion

So far we have considered morphologies mainly in primitive spectral and dynamic outlines, alluding to potential extensions through correspondence and stringing. In tackling the question of motion we begin to penetrate the intricacies of spectro-morphological design. We take it for granted that music is motion in time. As a result of electroacoustic music we have become increasingly conscious of an extensive range of types and patterns of motion. Musical motion has been made more apparent through the advent of stereophony, the continuing research into spatial localization, and the increasing sophistication of loudspeaker diffusion systems, bringing about actual spatial movement. Spectro-morphological design on its own, however, in controlling the spectral and dynamic shaping, creates real and imagined motions without the need for actual movement in space.

Motion typology is laid out in Figure 6. It is not suggested that musical motion, which may be a complex amalgamation of various types and tendencies, ambiguities and contradictions, should always fit neatly into one of the tabled categories. However, the five basic motion analogies represent the range of possibilities: *unidirectional*, *bi-directional*, *reciprocal*, *centric/cyclic*, and *eccentric/multi-directional*.

The motion categories may apply at a variety of structural levels and time-scales, from the shape of a brief sound-object to the motion of a large structure, from the groupings of objects to the groupings of larger structures. A motion category may refer to the external contouring of a *gesture*, or the internal behaviour of a *texture*. Imagine, for example, a structure whose external shape follows a divergent path but whose internal patterning is made up of particles which gradually conglomerate, or an accumulating structure whose internal components are small, parabolic objects.

All motion types have inherent orientations. Motion always implies a direction, however open, limited, ambiguous, minimal, or complex an implication or simultaneous implications may be. For example, a sustained, linear ascent is implicative because the line cannot continue indefinitely. It may fade into oblivion, it may reach a stable ceiling, it may attain a goal, or its

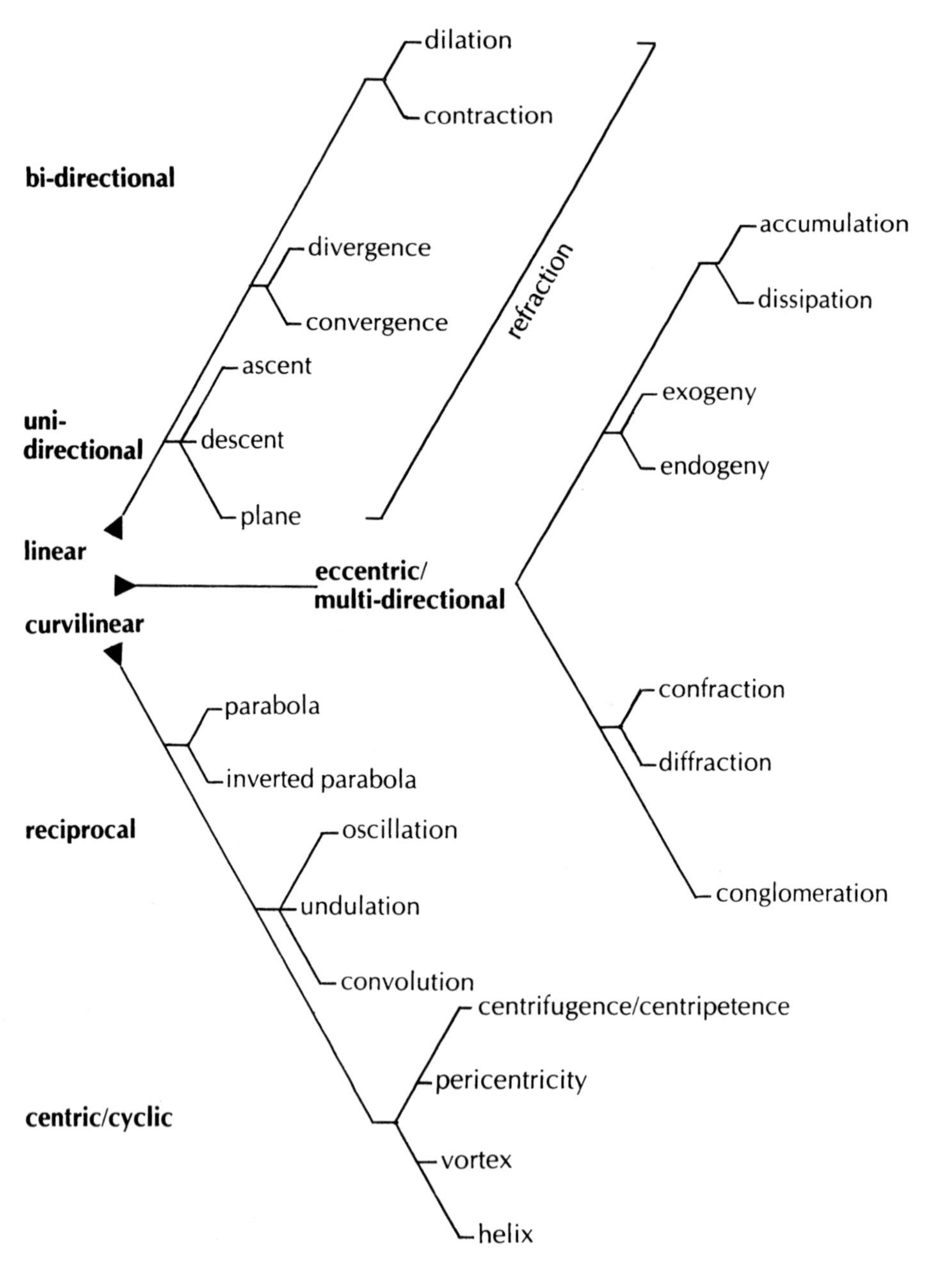

Figure 6: Motion typology

implied motion may be interrupted or excised by a new event. Spectro-morphological shaping and the musical context will provide clues to an implication or implications. In other words, spectro-morphological composition, like other musical languages, is concerned with expectations gratified and foiled, and such expectations are founded on shared perceptual experience. The failure to understand the directional implications and temporal pacing of motion is a common compositional problem.

The upper sector of the typology motion table expands the simple *linear* principle. The linear categories are self-explanatory, with the exception of *dilation* and *contraction* which remain linear only in a very generalized fashion. We can regard them as free translations of *divergence* and *convergence*. At their most wayward they could be thought of as eccentric motions. The incorporation of *refraction* reminds us that linear motion can be deflected off its course resulting in a change of angle or a change of direction. This may mean that linear motion transforms into curvilinear motion, particularly since a deflection is likely to be interpreted as a change in velocity and thus be translated into exponential terms. So linear and curvilinear distinctions are not clear-cut.

Curvilinear motion links three phases: ascent–peak–descent, or its inversion descent–trough–ascent. Curvilinearity is totally *reciprocal* if motion in one direction is balanced by reciprocation in the opposite direction. Partial reciprocation, where the third and first phases are different lengths, is common. Dynamic and spectral profiling, either separately or in collaboration, assist in articulating a variety of curvilinear shapes associated with changes in velocity. For example, a variety of thrown or bounced motions is achieved by appropriate variations in the rates and distances of ascent and descent, associated morphological profiling towards and away from the peak or trough, and careful proportioning of time taken to effect the change of direction. Finally, it seems that a sustained sound which changes direction is perceived as curvilinear. For a change of direction to be perceived as linear the motion and motion rate must be interrupted or accentuated at the peak or trough to create the impression of an angular direction change.

Oscillation, *undulation*, and *convolution* are extensions of curvilinear motion. *Oscillation* describes curvilinear alternation between two points; *undulation* implies shapes of varied dimensions tracing reciprocal motions; *convolution* implies a complex interweaving of reciprocal motion shapes.

Focus on a central reference point is implied in *centric motions*, either radiating from a centre, or converging on it. This does not mean that a central point has to be represented by an actual sound, just that we surmise the existence of a central focus from the surrounding motion tendencies. In visual terms centric motion is commonplace. But music, thought of in terms of time, and conditioned by notational practice, moves through successive stages in time, from left to right, thus seeming to prohibit any analogy with our visual scanning, which can shift backwards and forwards across a moving object. The musical solution lies in the cyclic nature of centric motions related to short-term memory. The recycling of motion allows us to re-perceive and therefore inculcate its turning shape and continuity in a series of film-like frames, and surmise centricity as a result. This assumes that the contours and morphological elements conspire sufficiently to create a centric impression in the first place. Careful morphological design aided by spatial articulation can yield very effective results.

It is not always easy to tie a centric motion down to a very specific type. The five centric terms suggest possible interpretative shades. *Centrifugal motion* expresses sounds flying off from the centre while *centripetal motion* tends towards the centre; *pericentral motion* means motion around a centre; *vortical motion* describes the movement of particles about an axis; *helical motion*, which follows a spiral shape, implies acceleration or deceleration and changes in dimensions during recycling. Finally, successive cycles do not have to be exact repetitions. Components and pacing can vary without destroying the centric focus.

Eccentric motion suggests a lack of central focus, but ambiguities are still possible. Centric motion may be constructed so as to incorporate elements of eccentricity, just as in eccentric motion centric principles may not be entirely absent, although the motion process is such that centric motions do not dominate. The companion term 'multi-directional' serves to emphasize the expansion of the linear idea via divergence and convergence which we noted can be developed towards eccentricity.

Exogenous motion describes growth by additions to the exterior of a sound, while *endogeny* denotes growth from within. The former implies an initial reference point while the latter implies a frame whose intervening space becomes increasingly active or dense. Both processes can work in reverse, and are specific categories of *accumulation*, a very general term, common both as a motion and as a structuring process. *Dissipation* is its opposite. *Confraction* describes a break-up into smaller fragments whereas *diffraction* implies break-up into a configuration of sound bands. *Conglomeration* expresses the reverse process, the forming of a compact mass or object from fragments or bands. All these terms can be related to spectral decomposition processes. Unlike other motion groups they are growth processes involving textural transformations whose morphological outcome is radically different from the initial state.

* * *

Figure 7 amplifies internal details of motion typology which we have called *motion style*. On the left of the diagram are the four basic continua which delineate the boundaries affecting the internal progress of motion: whether simultaneous components, objects or events act together or independently, whether or not motion proceeds in a continuous flow, whether contouring is evenly or unevenly graduated, and whether motion change is regular or not.

The central trio of terms identifies three typical categories of internal motion design, textures which may be made up of either a single morphological type (*monomorphological*) or a mixture of morphologies (*polymorphological*).

In *flocked motion* individual components behave in a coherent group or groups. Flocking can therefore only be monomorphological, but the ear

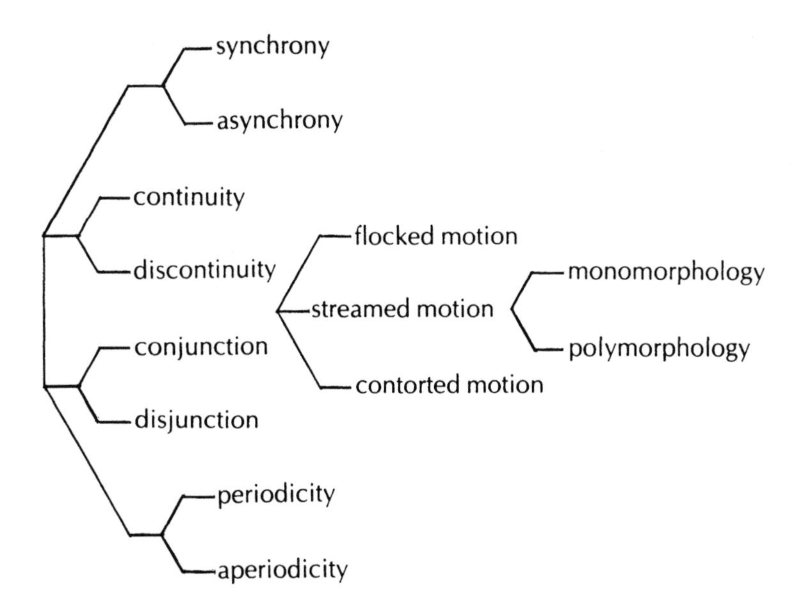

Figure 7: Motion style

follows the behaviour of the flock rather than individual components. The thinner the density or the slower the motion the more likely individuals will be to stand out. If flocking becomes more blurred or thicker in density we start to move through the attack–effluvium or pitch–effluvium continuum, eventually destroying the flock. Flocking therefore occupies the focal middle-ground between individuality and annihilation.

Streamed motion contains a concurrent flow of motions which maintain their separate identity. This may occur for a number of reasons: the streams could be morphologically different or possess familial similarities just sufficiently different to tell them apart; they could be contrasted in motion type; they may be separated out by textural gaps to form strata, but they need not move in parallel. *Streamed flocks* are possible (or *flocked streams*, depending on which is the more dominant factor). We can further qualify streamed motions by referring to the four basic continua in which we find possibilities for potential coincident relationships among streams, through periodicity or synchronicity, for example. Finally, it is extremely difficult for separate identity to be absolute since we immediately associate the streams through their concurrence, even though their patterning or behaviour may be drastically different. Streaming is therefore a brand of counterpoint.

Contortion suggests a relationship of components so entangled that they have to be considered as a whole. While individual components may leap to the ear they are nevertheless entrapped by a disorder around them which prevents streaming, and their erratic and overlapping contours will resist flocking. The nature of the chaos overwhelms its participants. Contortion is either polymorphological, or else the components are morphologically related, but sufficiently pliable, or extreme enough in their differences, to permit a contorted organization. In disorder there is a certain order, and contorted motion creates its coherence through a similarity of erratic behaviour shared by its components whatever their personality differences.

* * *

The above criteria concerned the internal motion of spectral texture. We need also to consider more stable textural contexts. In Figure 8 there are four basic reference textures called *stable pitch-space settings*. Tessitura must be linear and consistent, but stability does not necessarily require internal consistency of motion. Discontinuous or aperiodic motion when left long enough can also provide stable settings if the ear comes to accept the continuing existence of the motion rather than being drawn into its internal variations. Reciprocal

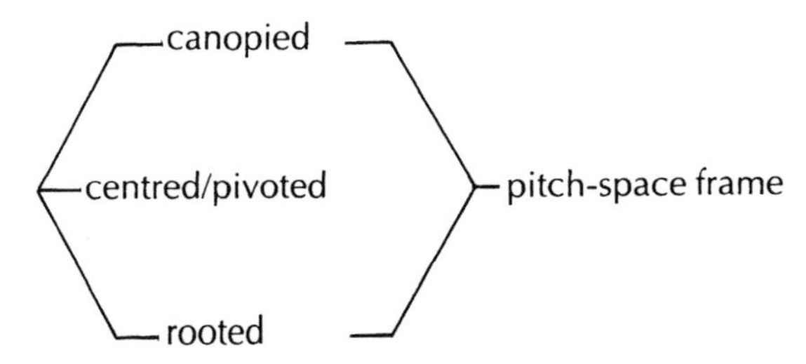

Figure 8: Stable pitch-space settings

and centric motions, if consistently patterned, may also provide stability. But the key word is 'setting' since it implies that other events may flourish within the pitch-space. The setting provides the reference point or boundaries of a pitch-space environment. If new or alternative events are introduced, our ear is drawn to them and we begin to take the setting for granted, pushing it into the background. This relationship is, of course, open to considerable variations of focal play.

In a *canopied setting* musical activity takes place beneath a high pitch area. Such activity may be independent of the canopy, suspended from it, or spectro-morphologically linked to it.

Its opposite is the *rooted setting*, to which the same comments apply. But the rooted setting also has traditional pedal-point connotations, which have continued in electroacoustic music in the guise of drones. Thus there may be associations with the idea of a fundamental tone out of which spectral motion grows. This is the classic example of the permanence of a note-base being taken for granted while the ear follows spectral variations above the root.

The canopy and root together form a *pitch-space frame*. Such a polar formation sets boundaries which can invite analogies with the experiencing of the dimensions and grandeur of visual space.

Finally, there is the *centred* or *pivoted setting* which provides a central reference or pivot around which musical events take place, from which events spring, or towards which events gravitate. The connections with centric motions are clear, but its central role, particularly if it is note-based, may mean it is less content to be relegated to the background, and it is therefore likely to move into the foreground.

* * *

We have now covered the basic features of spectro-morphological design, expanding notions of spectral typology into morphologies, motion typology, motion style, and stable pitch-space settings, which have inevitably introduced the idea of *spectral texture*: the vertical distribution of the constituent parts or components of pitch-space. We found that *fusion* and *diffusion* (or *dispersal*) are important notions affecting our perception of spectral texture. We may prefer to think in terms of textural *opacity* and *transparency*. Opacity is a state where effluvial criteria block out detail, and transparency opens up textural interstices which permit an impression of spatial clarity. We can think of components as occupying certain *sites* inside the boundaries of pitch-space, of each site as having a *dimension*, and of the spectral texture as perhaps being weighted in favour of one site or more at the expense of others. The idea of *spectral weighting* draws attention to the focus attributed to a particular component or component group. A complex interaction of factors may influence weighting: spectral type, morphology, dynamic prominence, whether the component is sandwiched between textural gaps, the dimensions of such interstices, in short, whichever contextual factors aid the diagnosis of separate identity.

Structuring processes

Level and focus

In spectro-morphological music there is no consistent low-level unit, and therefore no simple equivalent of the structure of tonal music with its neat hierarchical stratification from the note upwards, through motif, phrase, period, until we encompass the whole work. *Significant units* at the lowest level of structure can be of considerable temporal dimensions as we have implied in the discussion of motion. But a 'unit' or a self-contained 'object' is often difficult or impossible to perceive, particularly in continuous musical contexts which thrive on closely interlocked morphologies and motions. There is not even a consistent *density referrent*, a pulse which sets the pace at which we begin to apprehend a work or 'movement' of a work. Indeed, increased temporal flexibility is one of the newly evolved bonuses of musical experience. We have to be more prepared to vary the pacing and intensity of our listening according to the changing spectro-morphological cues uncovered as the motion unfolds.

While we may be unable to detect consistent structural hierarchies, we nevertheless demand that a structure be *multi-levelled*: we need to be offered the possibility of varying our perceptual focus throughout a range of levels during the listening process. Indeed, a work must possess this focal potential if it is to survive repeated hearings during which we seek not only the rewards of the previous hearing, but also fresh revelations. If we cannot find permanent hierarchical relationships throughout a work we shall nevertheless often uncover *fractured hierarchies* of varying temporal dimensions and varied numbers of strata as our ears scan the structure.

A crucial reason for the failure of many electroacoustic works may be the composer's inability to maintain control over the focal scanning of structural levels during the process of composition. Particularly in tape composition, because of the need for the constant repetition of sounds during the honing process, the composer is too easily tricked into perceiving microscopic details which will be missed by the listener. Furthermore, constant repetition quickly kills off a sound's freshness so that the composer's assessment of material becomes jaded. On the other hand, an over-concentration on the design of the higher levels of structure can all too easily lead to a work lacking in the lower-level detail so necessary for the rewards of repeated hearings. Thus a work in extreme and consistent high-level focus will rapidly achieve aural redundancy, while a work myopically conceived in low-level focus will be found wanting in structural catholicity.

While these problems are not the sole prerogative of electroacoustic composition, they do achieve new significance in this medium, firstly because the composer has the added burden of finding or creating materials from scratch, and secondly because the intervention of technology can become just as easily a hindrance as a help, and the problems in designing and using it an end in itself rather than a servant in the employ of music.

So we return to aural discrimination and perception as the supreme musical tools. It is not a scientific knowledge which is required but an experiential knowledge. The composer has to surmount all the preoccupations and distractions of the fabrication process to become the subject of his or her own musical experimentation – the universal listener, the surrogate for all listeners.

Gesture and texture

The terms *gesture* and *texture*[7] here represent two fundamental structuring strategies associated with multi-level focus and the experience of the temporal unfolding of structure. We are already familiar with the special position accorded to the idea of gesture as an almost tangible link with human activity, but although we have referred to texture and *spectral texture* in discussing motion we have yet fully to define the term.

Gesture is concerned with action directed away from a previous goal or towards a new goal; it is concerned with the application of energy and its consequences; it is synonymous with intervention, growth and progress, and is married to causality. If we do not know what caused the gesture, at least we can surmise from its energetic profile that it could have been caused, and its spectro-morphology will provide evidence of the nature of such a cause. Causality, actual or surmised, is related not only to the physical intervention of breath, hand, or fingers, but also to natural and engineered events, visual analogues, psychological experiences felt or mediated through language and paralanguage, indeed any occurrence which seems to provoke a consequence, or consequence which seems to have been provoked by an occurrence.

Texture, on the other hand, is concerned with internal behaviour patterning, energy directed inwards or reinjected, self-propagating; once instigated it is seemingly left to its own devices; instead of being provoked to act it merely continues behaving. Where gesture is interventionist, texture is laissez-faire; where gesture is occupied with growth and progress, texture is rapt in contemplation; where gesture presses forward, texture marks time; where gesture is carried by external shape, texture turns to internal activity; where gesture encourages higher-level focus, texture encourages lower-level focus. Texture finds its sources and connections in all like-minded activities experienced or observed as part of human existence.

Of course, all this is too simple, but before we disturb the simplicity we shall digress to consider the mechanisms whereby musical gesture and texture are linked to their sources.

Let us regard musical instruments and their sounding gestures as stand-ins for non-musical gestures. This is *first order surrogacy*, the traditional business of instrumental music. If that instrumental source is electroacoustically transformed but retains enough of its original identity it remains a first order surrogate. Through sound synthesis and more drastic signal processing, electroacoustic music has created the possibility of a new, *second order surrogacy* where gesture is surmised from the energetic profile but an actual instrumental cause cannot be known and does not exist. It cannot be verified by seeing the cause. Second order surrogacy therefore maintains its human links by sound alone, rather than by the use of an instrument which is an extension of the body. Beyond this second order we approach *remote surrogacy* where links with surmised causality are progressively loosened so that physical cause cannot be deduced and we thus enter the realms of psychological interpretation alone. Ours is thus the unique historical period of second order and remote musical surrogacies which have shifted the burden of listening away from direct physicality. Ironically, though, it is also the period of increasing importance for physically-based sources since through field recording composers have direct access to sounding activities not previously accessible as musical materials. Thus for the first time we also travel backwards from first order surrogacy to tap non-instrumental sources. We therefore have direct access to gesture including a range of vocal gestures not previously accepted as musical materials.

While we are very aware of texture in musical compositions, texture as a feature of spectro-morphological structuring is more influenced by textures found in environmental sounds, by the magnification of the internal textural details revealed through sound recording, and also by the visual observation of textures in either animate or inanimate visual phenomena whether natural or man-made. While gesture has its origins entirely in the human body, texture is based either on the spectro-morphological detail found in the first order surrogacy of gesture, or on objects and phenomena independent of the human body.

Are there limits to surrogacy? Is there a stage at which the dislocation from experience becomes intolerable? Our remarks concerning the tolerances of spectro-morphological design, the implications of motion pacing, and multi-level focus are all symptoms of an unviable, *dislocated surrogacy*. Many a listener's problem can be related either to the loss of tangibility created by the severance of direct gestural ties, or to the difficulties in comprehending the remoteness of new surrogacy.

The relationship between gesture and texture is more one of collaboration than antithesis. Gesture and texture commonly share the structural workload but not always in an egalitarian spirit. We may therefore refer to structures as either *gesture-carried* or *texture-carried*, depending on which is the more dominant partner.

In a *gesture-carried* context gesture dominates. If the gesture is strongly directed and swift moving, the ear is more likely to ride on the momentum generated rather than dwell on any textural niceties at the gesture's interior. With a less impetuous gesture we can imagine the possibility of enticement by internal textural detail which would create a certain equilibrium between gesture and texture. The ear recognizes that a gesture is in progress: the sense of directed motion remains, and can be temporarily taken for granted while the ear shifts focus to delve into textural motion, perhaps to emerge again once a more urgent sense of directed motion is detected. We can regard these circumstances as an example of *gesture-framing*.

The more gesture is stretched in time the more the ear is drawn to expect textural focus. The balance tips in texture's favour as the ear is distanced from memories of causality, and protected from desires of resolution as it turns inwards to contemplate textural criteria. Gestural events or objects can easily be introduced into textures or arise from them. This would be an example of *texture-setting*, where texture provides the environment for gestural activities.

Both *gesture-framing* and *texture-setting* are cases of an equilibrium capable of being tipped in either direction by the ear. They indicate yet again areas of border tolerances, ambiguities open to double interpretation and perceptual cross-fadings depending on listening attitude. On the whole, gestural activity is more easily apprehended and remembered because of the compactness of its coherence. Textural appreciation requires a more active aural scanning and is therefore a more elusive aural art. Finally, we should not forget that attack–effluvium and pitch–effluvium tolerances are critical in determining the potential for gestural or textural apprehension.

Structural functions

We now resurrect the three linked temporal phases of morphological design – onset, continuant, and termination – as models for structural functions. In this way we project morphological design into structural design, thereby uncovering the links between low and high levels of structure. In the discussion of gesture and texture we have already almost unconsciously attributed these morphologically-based functions. In the directed motion of gesture the continuity of the three temporal phases was in evidence, or rather, an emphasis on action directed away from the onset, and action directed towards a termination or second onset. Texture, on the other hand, is based on the continuant model.

Figure 9 shows how the three temporal phases may be expanded to denote structural functions at higher levels. The functions are collected into three groups, each group covering a range of terms with common attributes. No doubt the terms can be expanded to incorporate different shades of meaning appropriate to their own interpretation of functional circumstances.

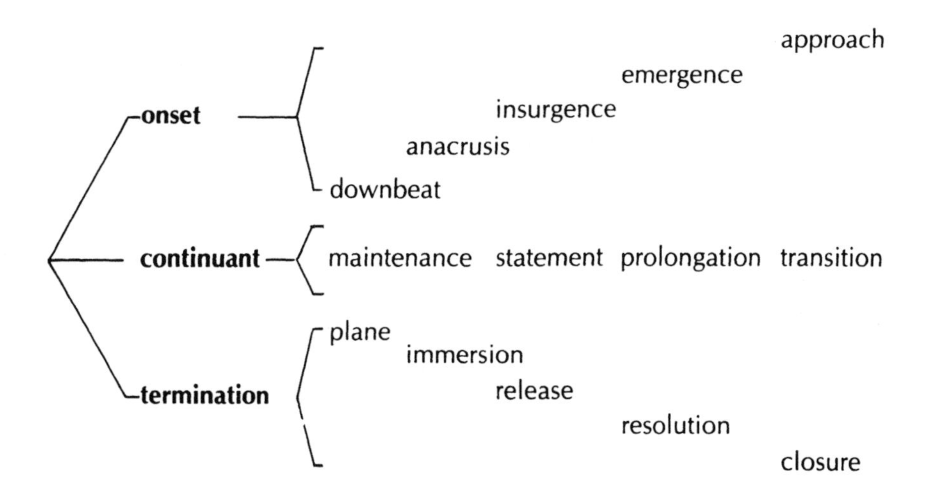

Figure 9: Structural functions

The *onset group* covers a range of possibilities, from the more specific and traditional *downbeat* and *anacrusis* to the less specific *emergence* and *approach*. The *continuant group* covers a broader variety as befits the ambivalence of its status: the term *maintenance* is fairly non-committal, merely noting that whatever has been set in course continues; *statement* is more definitive and self-important, reminding us of the traditional 'exposition'; *prolongation* expresses functional stretching, a lying-in-wait for more definite continuity; and *transition* describes the intermediate stage between other supposedly more important functions. We therefore realize that the continuant function is not neutral: time cannot stand still, and real stasis is not possible. In the *termination group*, *plane* is the most ambiguous. A plane is felt to be a zone of arrival or a lacuna, a relatively stable state interpreted as a goal of what has come before. The remaining terms have their counterparts in the onset group, except for the inexpressive word *closure*.

We may attribute functions at a variety of structural levels depending on the aural focus operative at the time, and the stage reached in the unveiling of the total structure. To begin with, our interpretation of structural function will operate on a *local* scale, but as the music progresses more is revealed so that we inevitably interpret functions in the context of more *regional* circumstances. When the work is complete we have the opportunity to reassess its structure in the light of our experience of the whole. The attribution of structural function as a result of short-range interpretations may become invalid once the wider context is revealed. Moreover, a superposition of functions relating to different temporal dimensions occurs because functional attribution is carried out at all levels of structure. For example, a localized object may be part of a larger grouping which is ascribed quite a different regional function. Once the region is experienced as part of a larger grouping a third attribution may be ascribed to this more *global* context. The attribution of structural function is therefore a hierarchical game, or more accurately, a multi-level process. Thus, during the passing present we are for ever weighing up, altering and superimposing our interpretations of function. What we interpret depends on our aural acuity, how good our aural memory is, how we unconsciously or consciously decide to focus our aural scanning, and of course, the skill with which the composer has prepared the musical structure for our apprehension.

However, interpretation of function is not necessarily a decisive process. Listeners and structures thrive on ambiguities. During the act of listening more than one function can be simultaneously attributed to a single level of musical activity. These are interim attributions subject to confirmation or change once the present has passed on and potential musical implications are realized or foiled. But a single, definitive attribution is neither necessary nor always possible, particularly if the structuring involves continuous motion and tight interlocking of events. Insecurity is part of the musical process and we can be quite happy to be left with dual or even multiple attributions which reflect our experiencing of functional ambiguities.

Let us simulate the interpretative process by pursuing two hypothetical *function chains* which might occur in the course of a work:

> anacrusis→statement?/prolongation?/transition?/plane? =
> plane = closure

This structure starts with an anacrusis and moves into the continuant phase which solicits a series of functional queries. Is this a major statement of significant material? Is it merely a prolongation of the anacrusis? Are we entering a transitional stage, or as the continuant phase is prolonged are we experiencing a plane? We finally attribute a planar function which we discover collaborating with a closing function. We choose the more neutral 'closure' because we are slightly dissatisfied with the commitment of the terminal phase and hope that this will not be the final termination of the work.

The second example occurs in three stages:

1. initiation/statement? = initiation/statement
2. statement = plane/transition? = transition
3. transition→closure? = immersion/insurgence/emergence

The beginning of the structure (the onset-initiation) and the continuant statement are dual attributions. Right at the outset we diagnose the structural importance of the statement function. As the statement continues we begin to invest both planar and transitional qualities in the context but finally settle for the transition, which we surmise as heading for closure. But the closure turns out to be the more committed immersion occurring simultaneously with two onset functions, the more impetuous insurgence, and the more careful emergence. Eventually, therefore, we prefer not to separate the interwoven functions of initiation, statement and transition, and we definitely cannot separate out the triple immersion, insurgence and emergence. There is no time for a real feeling of resolution, and any semblance of termination is negated by the superimposed onsets.

Every function does not have the same significance. There must be a variety of *function weighting* in a musical structure to create satisfactory balancing of temporal stresses. Excessive repetition of the same function configuration, for example, will court rapid aural indifference even if the pacing is varied, simply because the stress-patterning is too predictable. One may also confront works which are so limply articulated that functions can hardly be attributed, or where seemingly endless ambiguities induce psychological paralysis. There are works so preoccupied with inner articulations that there are no regional or global functions to be found, as a result of which the need for multi-level focus and high-level coherence is disdainfully dismissed. The litany of reasons for structural collapses is lengthy.

The success of a musical work rests on an instinctive interpretation of function balancing by the listener. Furthermore, it has been argued in these discussions that any dissatisfaction with structural functioning can be traced back through the formative notions of gesture and texture to the spectro-morphological sources from which structural functions are projected. The triple-group concept of structural functions is not the invention of electro-acoustic music. It is inherent in our experiencing of time passing. Even if in the extreme bifurcation of modern musical language spectro-morphology seems to sweep aside traditional musical notions, the foundations of structural functioning remain.

Structural relationships

Figure 10 shows the relationship of simultaneous and successive structural components, indicating how events which are perceptually differentiated at a variety of structural levels act together. All the terms may be used as useful supplements to describe the internal workings of motion, gesture and texture, gesture-framing and texture-setting. They refer primarily to local and regional environments.

True independence is not a musical reality. It is rare if not impossible for simultaneously existing events to be unrelated, simply because placing them together in a musical context confers connection upon them. That connection is forged from one of three directions. The first is concerned with *interaction*

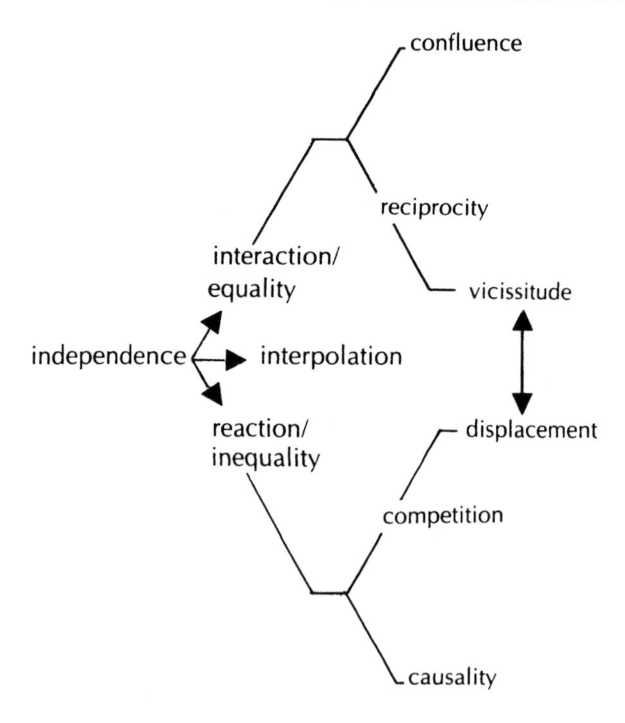

Figure 10: Structural relationships

or *relative equality*, the second with *reaction* or *relative inequality*, and the third – *interpolation* – most closely approaches independence.

Interaction means cooperation and is represented by *confluence* and *reciprocity*. *Reaction* implies either a *casual* or *competitive* relationship, either of which may involve degrees of active–passive role-playing. *Vicissitude* and *displacement* represent related methods of progression. In a *vicissitudinous progression* one event will willingly give way to the next. Such would be the relationship in a texture which is subject to gradual and continuous morphological transformation. *Progression by displacement*, on the other hand, expresses an event's resistance to being replaced. *Interpolation* is either interruption or change at a stroke.

Form

Has electroacoustic music uncovered any new forms? We have studiously avoided the word 'form' which historically has come to imply a relative consistency of external structural design common to several or many composers. Instead we have adopted the terms 'structure' and 'structuring process' applying them to both low- and high-level units and dimensions of the musical work. They idea of any fixed formal consensus is inimical to the unprecedented diversity of spectro-morphological materials. Moreover, the older idea of 'musical form' is based on the fixed and measurable cardinal virtues of pitch and metrical time. Spectro-morphology is not based on quantities, but on the perception of qualities whose complex nature resists the permanent or semi-permanent systematizing necessary as a foundation for formal consensus. So it is better to avoid the confusion with traditional notions of what form is by not using the term. However, we have found that there must be a consensus about structuring processes if spectro-morphological music is to find understanding among its listeners. On these natural foundations a proliferation of new and unique structural configurations has been composed.

Space

Space has been left until last firstly because a prior knowledge of the whole range of our discussions is needed, and secondly because it leads to the circumstances in which electroacoustic music is heard and interpreted by the listener. Space is a topic whose voluminous capacity can only be touched on in this essay. There are five dimensions to the consideration of space: *spectral space*, *time as space*, *resonance*, *spatial articulation* in composition, and the *transference of composed spatial articulation* into the listening environment.

We have already noted the propensity of spectral texture to suggest spatial analogies. It seems that we react adversely to any continuing tessitura restrictions of spectral space. A certain *equilibrium* of spectral spread is expected in the course of a work. *Spectral cramping* is a common deficiency in electroacoustic music and is often directly attributable to inadequate sound quality.

Spectral motion is perceived in time, and the ability of spectro-morphological design to create real and imagined spatial motions without recourse to actual spatial movement has been pointed out. Spectral space and time as space are the first two aspects of space common to all music.

The third aspect, *resonance*, is also universal. Resonators have always been integrated into instrumental and vocal morphologies. They are responsible for ensuring a sound's continuation beyond the onset phase, and enable the acoustic projection of spectral qualities. Resonance, then, is an omnipresent morphological property. It is an enclosed, *inner space*, a causal determinant of spectral behaviour. In electroacoustic composition we can harness these inner spaces to develop new morphologies. We can stretch out new, imagined resonances creating structures whose textural coherence retains the internal logic of resonance instigation, but may expand into fantasy. These are *resonance structures*, hybrids of gesture and texture, expanded from the second morphological archetype.

Spatial articulation in composition is the child of electroacoustic music. If we regard resonance as the inner space, then spatial articulation is the *outer space* where a sound structure interacts with the properties of the acoustic environment it inhabits. This is resonance at the exterior of a morphology, otherwise known as reverberation.

The relevant properties of spatial articulation are set out in Figure 11. We recognize a *spatial setting* which possesses *dimensions* either *confined* by reflecting surfaces or left more *open* as in the environment. A *realistic space* is a plausible setting while a *fictitious space* or change of spaces could not exist in reality. The setting is deduced from spatial behaviour which is able to encompass an immense repertory of motion, but not independently of spectro-morphological design and motion.

There are two common strategies, both directly related to the collaborative

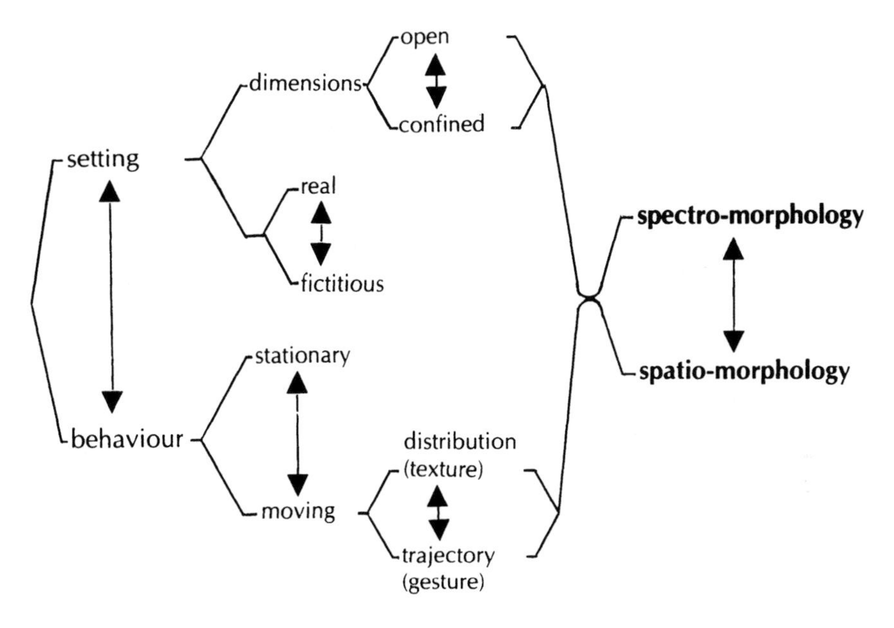

Figure 11: Spatial articulation

concepts of gesture and texture. Gesture is reflected in *spatial trajectories*, and texture in the *spatial distribution* of components. The perception of structural functions may be clarified, delineated, enhanced or obscured through the combined functions of trajectory and spatial texture. Thus spatial articulation acts as an arbiter of spectro-morphological and structural articulation. A suitable space is created to articulate the musical structure.

But the interaction of sound-structure and space can result in morphological transformation. The consequences will be perceived through changes in spectral richness and clarity due to distancing and velocity, and perhaps also changes in spectral contour if velocity is rapid enough to cause Doppler effects. Thus spatial articulation can become a morphological determinant.

A third situation is possible. It could be that a morphology is set in a sequence of spaces such that no significant morphological change occurs. We diagnose a change of spatial perspective through changes in the reflective properties and perceive a transformation of space rather than morphology. In this way the aural interpretation of space attains a new structural status. This new musical element will be named *spatio-morphology*, and in Figure 11 note its musical interaction with spectro-morphology in a continuum which at one extreme focusses on spatial information gleaned through spectro-morphology, and at the other exhibits a more frequent tendency to focus on spectro-morphological information.

The fifth spatial aspect is involved in the listening process where the music is transferred via loudspeakers into a new acoustic space – the listening environment. Neither the electroacoustic means of transferral nor the final space is neutral: both affect musical substance and structure. In performance it is this fifth aspect which may make or break a musical structure. This last act of adaptation and spatial interaction involves a reinterpretation of the musical structure for the listener so that it behaves in a manner appropriate to the dimensions and acoustic properties of the final space. It is a question of adapting gesture and texture so that multi-level focus is possible for as many listeners as possible, and if the space eats away at low-level detail then countermeasures must be taken. This is the fragile art of *sound diffusion*. In a medium which relies on the observation and discrimination of qualitative differences, where spectral criteria are so much the product of sound quality, this final act becomes the most crucial of all.

Conclusion

There are two facets of electroacoustic music which would have to be examined for a more complete perspective of the medium. These are *language* and *mimesis*. The use of words in electroacoustic music has been ignored in this chapter because it involves a separate investigation prior to considering its incorporation into music; and beyond noting how inextricably music is bound up with mimesis, we have left discussion of this topic to others, commenting only that any extra-musical message conveyed in a strongly mimetic work is carried and articulated by spectro-morphology.

The range of sound sources available as materials for electroacoustic music, and the scope of spectro-morphology's influence demonstrate an unprecedented broadening of our conception of the nature of music, demanding of the composer a much deeper and broader understanding of the role of sound in human life, pointing out to those in many other disciplines that musical concerns invade their area of influence, indicating to the musicologist the need for interdisciplinary knowledge and research, and inviting the listener to participate more fully and thoughtfully in the sounds of the world:

music permeates life like no other art. Perhaps as a result of electroacoustic music's preoccupations we can envisage a new, more universal breed of musician whose skills and insights could re-create for music the central cultural position it once maintained.

In conclusion, we claim that the very rapid development of spectro-morphology is the most radical change in Western musical history. In less than fifty years the materials of music have changed utterly, and we must now realize that spectro-morphological thinking is the rightful heir of Western musical tradition.[8] Spectro-morphology reaffirms the primacy of aural perception which has been so heinously ignored in the recent past, and warns composers, researchers and technologists that unless aural judgment is permitted to triumph over technology, electroacoustic music will attract deserved condemnation.

Problems of Language

5
At the Threshold of an Aesthetic

David Keane

Introduction

This chapter is concerned with a search for the basis of what might be called the aesthetics of electroacoustic music. The obvious starting place is the question: "What is electroacoustic music?", to which one might offer another question in reply: "Is it not first and foremost a subset of music, and would not the basis be found in the aesthetics of music?".

It would be much easier for us if we could begin our search by determining what electroacoustic music has in common with music in general, but as recently as the Third Colloquium of the Confederation Internationale de Musique Electroacoustique in Stockholm, one of the world's finest and most senior composers of electroacoustic music, Jan W. Morthenson, argued that "Electronic music is not rejected [by the general public] as meaningless because ordinary listeners are reactionary or insensitive, but because it cannot, under present cultural conditions . . . be psychologically classified as 'music'."[1] Morthenson's position is not an isolated one. A recently formed organization in Canada has called itself the 'Canadian Electroacoustic Arts Alliance', carefully and consciously eschewing the word 'music'. To begin with an assumption with which many would disagree would simply discredit this investigation into the nature of electroacoustic music before it began.

If we cannot assume that electroacoustic music is music, we have to retrace our steps in order to determine what we *can* assume it is. During the opening decades of the twentieth century, the telharmonium, the theremin, the ondes martenot and other electroacoustic instruments were introduced into a musical world that was struggling with the question: "What is music?". Of course, the musical world had struggled with *that* question since at least classical antiquity, but, by the second quarter of this century, the work of such composers as Satie, Russolo and Schoenberg had shaken the foundations of music theory in a way that went far beyond that with which other ages had had to contend.

The new electroacoustic musical instruments were certainly considered radical, but these did not in themselves challenge the traditional concept of music to anything approaching the degree that the non-electroacoustic innovations of some postwar composers did. Indeed, the early electroacoustic instruments relied *more* upon classical notions of music than upon what was then happening at the frontiers of contemporary music. However, we see a different phenomenon by the middle of the century, when composers like Pierre Schaeffer and Karlheinz Stockhausen appropriated powerful devices, such as the tape recorder and the tone generator, which were not originally intended for use in musical performance. The new electroacoustic instruments were not limited in the sounds that they could produce to those which are traditionally termed 'musical' (as is, for example, the violin). The tape recorder and the tone generator certainly had finite possibilities, and these possibilities and limitations have not been to any significant degree influenced by what past ages have seen as musical needs or potentials. Electroacoustic devices offered opportunities not only for the radically expanding musical expression of the mid-twentieth century, but also for disconnecting from musical tradition altogether.

In the 1950s and the years following, the tape recorder and the tone generator were taken up by composers eager to start afresh and push the frontiers of music ever outwards. However, that nagging question "What is music?" remained unanswered. From the late 1950s onwards, the designs for voltage-controlled synthesizers and digital systems have been based in part on assorted components of musical tradition, such as piano keyboards and standard musical notation, and in part on hardware peculiar to the audio technician's test bench, such as oscillators and filters. Another element in the design comes from the early experience with tape recorders and tone generators. For example, the dominating interest in serial structures in music stimulated the development of the sequencer, which in turn greatly influenced the structure of electroacoustic works in the early days of synthesizers.

In the centuries, even millenia, prior to our century, musical instruments and musical stimuli were endlessly formed, tested, and reformed. In this empirical manner, for example, the violin achieved, centuries ago, its present degree of perfection. Before assuming the form that we now know, the violin was defined and redefined by luthiers initially drawing upon experience of earlier instruments. With time, the roughness of 'misdirection' is identified and the course of musical development is smoothed and polished, or at least given that appearance. Consequently, it is no simple task for a few people in a few days, or years, to overthrow empirical tradition and attempt to create musical instruments or musical structures which will function as well as the traditional ones, or which will function well at all. Thus, we cannot assume that the use or apparent potential of electroacoustic musical instruments is necessarily an embodiment of the aesthetics of the medium.

In the case of the violin, a present-day composer may assume, and the listener may expect, certain predetermined limitations of timbre, range, and loudness as well as physical limitations to the ways in which a human player may incite to action such possibilities as the violin has. But such electroacoustic resources for music as the tape recorder and the tone generator came about in a matter of years, without reference to musical traditions and (because the resultant sonic energy does not necessarily correspond to actual or imagined human energies) without reference to human gesture[2]. The things that the electroacoustic devices could do, and could not do, very rapidly became a part of contemporary musical practice and this development took place with little reflection on the part of composers. Processes not particularly suited, even alien, to musical appreciation or perception found a place in electroacoustic music. One example of this was the degree of complexity that became part of some electroacoustic music. The youthfulness of the medium and its potential to extend well beyond the physical or psychophysical bounds of musical and artistic tradition therefore requires special scrutiny when we reflect upon the appropriateness of electroacoustic music-making today. On the other hand, perhaps we are not considering a resolution into the mainstream of music at all; perhaps it is at this point that music and electroacoustic music go their separate ways.

Although many of the issues raised in this chapter might be seen as the common concern of many forms of contemporary music and of contemporary art in general, the special circumstances in which electroacoustic musical creation takes place makes these issues particularly relevant to the medium. The broader issues which are shared among the arts in general seem to be the ones least considered in the examination of the aesthetic basis of electroacoustic music, yet these are the very ones that must be considered first.

One of the most important of these broader considerations is an understanding of the *purpose* of electroacoustic music. We are certainly better able to comprehend the nature of something if we know its function. One of the more likely functions of electroacoustic music is to establish communication between composer–performer and audience. But, if this is the case, it is complicated by the fact that on the one hand the general public seems to regard *contemporary art* with contempt, and on the other hand there are artists who regard the *public* with contempt (and who produce art for some more deserving public which will, they assume, exist in a better time or place).

For its part the wider public more often responds not to contemporary art itself, but to the superficial art commentary that surfaces in the popular press. A quick glance at some not atypical titles of articles which appeared in the English language during the formative years of electroacoustic music is suggestive:

'Composing Music by the Yard', Bennet Ludden in *Musical Courier*, January 1954
'Music Untouched by Human Hands', Roger Maren in *Science Digest*, August 1957
'Music without Musicians', Joel Tall in *Saturday Review*, 26 January 1957
'Madcap Music Machines', F. L. Remington in *Coronet*, March 1959
'Twilight of the Musicians', W. J. Paisley in *High Fidelity Stereo Review*, August 1960
'The Machine Closes in', *Time*, 16 February 1962

These titles are as misleading as they are threatening and they are accompanied by equally misleading and threatening articles. Although a significant portion of today's public was young or unborn at the time these articles appeared, attitudes, like musical traditions, linger – and, perhaps owing to the public's conservatism where the arts are concerned, attitudes about the arts seem particularly to linger. Although electroacoustic music is no longer reviled so aggressively the scars remain. There is no area of music that does not rely at times upon electroacoustic techniques, yet the general public's reaction to anything *called* electroacoustic music is often quite negative.

Many of the composers and performers of electroacoustic music have the idea that they must strive to free themselves from the suffocating tradition which the public seems to champion. It is not the tradition of art that suffocates, however, but the narrowness of the language of theory and aesthetics. Were the language of theory and aesthetics able to present to both public and artists a clearer insight into the nature of art, the hostility would be likely to dissolve on both sides. In the following discussion we will work towards a clearer insight into electroacoustic music. Our search for a basis for the 'aesthetics of electroacoustic music' will take us through consideration of three basic topics: the language of electroacoustic music; the art of electroacoustic music; and the instrumentality of electroacoustic music.

The language of electroacoustic music

We will begin by looking at the nature of language because the title of this book proposes that among the potentials of electroacoustic music are the attributes of language; or that electroacoustic music is itself a language. Neither proposal is likely to startle many readers: speculations about the myriad forms and places in which the elements of language are to be found have frequently been put forward since Noam Chomsky, in his *Syntactic Structures*[3], presented scholars with a lever that promised to be particularly useful in prying open the doors of knowledge.

Linguist Benjamin Lee Whorf[4] claimed that language is a means of interacting with the world and that each language guides its culture to perceive the world differently. If language is a channel for individuals to interact with the world, why should electroacoustic music not be thought of as a language? Electroacoustic music is increasingly an influence on pop music, advertising, film music and home entertainment. Moreover, there is precedent from recent decades for looking upon art, and music in particular, as language.[5]

Thus the idea that the word 'language' can be usefully applied in a profitable way to art and music is well established. If electroacoustic music *is* a language, or at least if music and natural language have sufficient features in common to justify the metaphor, we might be able to apply what we know of language to reveal otherwise obscure information about the relatively young field of technological music – we can apply the linguistic lever with assurance. The motivation for making any analogy is to see in what ways two entirely separate parts of our experience have elements in common *and* to see where the common ground ends. The point at which the comparison breaks down may be as informative (if not more so) as the points which are in common.

It is well known that Chomsky saw in each language a *deep structure* of the basic patterns and concepts of how the language works, and several or many *surface structures* or actual sentences, generated and transformed from the deep structure. The idea of a parallel arrangement of theoretical form and real-world incarnations has been attractive to scholars trying out such dichotomy in mathematics (in which Chomsky's work has its roots), art, communications, social structures, and many other aspects of human endeavour. Chomsky's avowed purpose was to frame a theory to account for the inexhaustible freshness and originality of language – its potential for creative genius. Campbell puts Chomsky's central view succinctly:

> Language does not wear meaning on its sleeve. Underneath a spoken sentence [surface structure] lie abstract structures [deep structure], to which the sentence itself is related only indirectly. These structures are, in a sense, plans or descriptions of sentences. They exist on more than one level, and are related to one another by means of rules. If the meaning of a sentence at one level is not explicit, it will be explicit at another level, where the parts of the sentence may be in a different order. To make clear the precise relationship between parts of a sentence, as intended by the speaker, both levels of structure are needed. This is the only way to make sense of many conversational utterances. A listener is able to grasp this relationship, because his own competence gives him access to the rules which connect the surface performance version with the other versions underlying it.[6]

The acid test, then, for our metaphor is the issue of 'competence' in Chomsky's theory. We might well imagine an electroacoustic musical 'sentence' and an abstraction of that 'sentence', but we run into significant difficulty when we tackle competence. Competence is the tacit knowledge possessed by a native speaker of all (and only) the well-formed sentences of his own language. The speaker has this knowledge, and if a listener is to benefit from the gesture, the listener shares that competence. During the communication the listener may have to assess, for example, what language is being spoken and what is implied by the way in which the language is being used. The basic rules of how the chosen language works, however, are already in place.

Jos Kunst indicates that music does not rely on this same kind of competence:

> We would be wise to avoid some pitfalls and dead ends that more than one has fallen victim to. I refer to the too-easy equation of musical activities to those within natural language processes. One should be warned off already by the fact that at a very early stage of human development the learning of natural language is, in an important sense, completed, whereas it seems that no such thing can be said of things like poetry or music.[7]

There is a very famous statement made by Bertrand Russell: "Understanding language is . . . a matter of habits acquired in oneself and rightly presumed in others."[8] Can a specifically electroacoustic musical competence have been acquired by its 'speakers' and 'listeners' in the short span of decades since its advent? If a basis for competence in an electroacoustical musical language exists, it exists at an extremely primitive level.

Even if the electroacoustic music-language metaphor is interpreted more loosely so that it is the element of *communication* in language, rather than its syntactic function, that is stressed, it will take us only a moment's reflection to realize that the problem of competence is not resolved. Even before the advent of communications theory, George Mead wrote: "Communication takes place only where the gesture made has the same meaning for the individual who makes it that it has for the individual who responds to it."[9] And at the dawn of communications theory Gilbert King postulated: "Every system of communication presupposes . . . that the sender and the receiver have agreed upon a certain set of probable messages".[10] The word, 'communication' derives from *communicare*, a Latin word which means literally: 'to make common'. We find ourselves back at the same place: language and language-like operations require a common set of assumptions among the users of the system.

Music has often been described by the scholarly and the not-so-scholarly as a 'universal language', related perhaps to the concept of universal grammar elaborated by Chomsky. If so, the limits that define this universal grammar are simply and obviously the limits of human perception and the human ability to interpret and respond. But now we are no longer dealing with something specific to language or communications; we are considering the dimensions of human perception, human knowledge, and human power to act upon the world. Language and electroacoustic music have something in common, but that something is the same thing that *every* aspect of human behaviour has in common. We have said that there may be something to learn at the point where a metaphor breaks down, but if the metaphor fails at such an early point that we are entirely unprepared to observe it, that metaphor may become seriously misleading. Language is very powerful. Once we put a label on something we sometimes begin to respond to the label in the place of the thing that was labelled. Postman and Weingartner warn that:

> The structure of our language is relentless in forcing upon us 'thing' conceptions. In English, we can transform any process or relationship into a 'thing' by the simple expedient of naming 'it' into a noun. We have done this with 'rain' and 'explosions', with 'waves' and 'clouds,' with 'thought' and 'life.'[11]

For this reason, Chomsky has recently dropped the term 'deep structure'. He has done so because people mistakenly read into it that language at this level is literally more profound than at the 'surface' level.

When people say apologetically that they don't know anything about music it is quite often the case that they have a very broad and deep experience of listening to music, with strong likes and dislikes, and in some cases a considerable ability to reproduce music previously heard or to improvise new music. What they really mean is that they don't read musical *notation* or that they know insufficient *words* about music. Our predominantly symbolic way of communicating musical process has so distorted our general perception of what 'knowing' about music is, that somebody with a command of everything that matters in 'knowing' about music can be made to feel ignorant. Meanwhile the people who have charge of the symbols and words remain reasonably complacent that they do 'know' about music.

We have substituted words for music for such a long time that we have quietly acquired the idea that in an important sense music and language are the same thing. The idea is so well rooted in our thinking that in the second half of this century we began to try to understand the nature of music by examining the nature of words. That is to say that we looked to *linguistics* to replace the concepts of music theory that had failed to give us the information we needed to create a functional aesthetic examination of contemporary music. Perhaps we are coming to understand more about the words we use *about* music through looking at linguistics, but we have not moved much closer to an understanding of music. Nevertheless, let us not yet abandon our look at language; let us look a little more broadly.

As far as the listener is concerned language is largely concerned with the questions: "Who or what is the source of what is being communicated?", "Why is this being said?" and "What does the communication say or mean?". While the first two questions are certainly important in understanding the communication, they are supplemental to the central issue of the third. There is a fundamental difference where music is concerned. If music can say or mean anything it is entirely in the domain of the first two questions. A musical expression cannot be paraphrased, condensed, or elaborated with regard to meaning. A musical idea is only what it seems to be. *Being* is its essence. Music does not stand for something else, but is concerned with the way experience itself progresses.

The art of electroacoustic music

Poetry has often been thought to mediate between language and music. Compared to prose, poetry lends itself less to linguistic interpretation and more to sheer sensory experience. Kunst continues his argument against the language metaphor by saying that indeed, the parallel with poetry, as far as it goes, seems very instructive with regard to music. Although poetic practice is based on natural language competence, it can actually run counter to it. Rhythms may resist the natural rhythms of speech; rhyme begins to undermine the (as Saussure describes it) 'arbitrariness of the sign' in relation to what it signifies, and the 'form of expression' becomes entangled in the 'form of content'.[12]

Poetry is art fashioned from the materials of language. *The Harper Handbook to Literature* states:

> As more or less rigorously defined by various writers [such as Poe and later the French symbolists including Baudelaire, Mallarmé, Verlaine, Rimbaud, and Valéry], pure poetry excludes narration, didacticism, ideas, and other elements of content that could be expressed in prose. Essentially lyric, the pure poem stands apart in its poetic essence, an object of beauty, superbly crafted, non-referential. Frequently, an analogy is made to music, with the emphasis on the poem as a pattern of sounds removed from their prose meanings.[13]

The practicality of creating pure poetry is not our concern here; the point is that although poetry uses language, we find poets striving to understand and exploit what one might call the 'music of language'. There is something in music that may also be found in language, but something that is *independent* of the functions with which language most concerns itself. Language is a vehicle for content, but pure poetry eschews content and concerns itself with the intrinsic value of pure structure.

Although there is no evidence that electroacoustic music is any more like language than other music or other arts, there would appear to be a common ground which might well relate to the abstract element which the pure poets identified as being in music. Rather than trying to understand electroacoustic music as a language, we would do better to search for an understanding of electroacoustic music in the context of this common ground. Moreover, if we have learned anything from our examination of the nature of language, it is that we should choose our words carefully, and that we should never underestimate the influence of words over how we perceive what we are describing. While we may question Morthenson's conclusion that electroacoustic music ought to be thought of in a different classification from music, he seems rightly to have identified the muddle that we have got into in trying to talk about electroacoustic music using the framework of conventional music theory and aesthetics. The problem may not be the nature of music, *per se*, but the nature of the *conventional* framework. Morthenson seeks to clarify the issue by rejecting not only the word 'music' but also 'concert', and 'composer', substituting for them 'audio art', 'audio exhibition' and 'audio artist'. He proposes that "a new set of basic psychological and aesthetic components . . . be established" to offset the very strong habits created in the old musical thinking.[14]

Morthenson offers no evidence that 'audio art' is actually a more satisfactory way of representing the activity that concerns us, but let us give the terms the benefit of the doubt for a time, and go on to consider whether a new set of basic psychological components *are* necessary for the foundation of the aesthetics of electroacoustic music/audio art. There is a clear indication that we need to start at a more basic level than the conventions that language assumes and, in any case, an examination of auditory perception seems as appropriate a starting point for an examination of music in general as for electroacoustic music. Perhaps we may be able to identify not only what language, music and audio art have in common, but also the point at which music and audio art take separate perceptual courses (if indeed that should be the case).

Perception and electroacoustic music

We cannot make a better beginning than to heed Leonard B. Meyer's warning that:

> ... while it is commendable for composers to be concerned with the limitations of the senses, it is well to remember that music is directed, not *to* the senses, but through the senses and *to the mind*. And it might be well if more serious attention were paid to the capacity, behavior, and abilities of the human mind.[15]

There is a great deal written about what art is, but one of the most succinct statements about the relationship of art to the function of the mind is that of the Russian structuralist, Victor Shklovsky:

> Art exists to help us recover the sensation of life; it exists to make us feel things, to make the stone *stony*. The end of art is to give a sensation of the object as seen, not as recognized. The technique of art is to make things 'unfamiliar,' to make forms obscure, so as to increase the difficulty and the duration of perception. The act of perception in art is an end in itself and must be prolonged.[16]

What we want art to do then, is to intensify the act of perception beyond the ordinary operations of the mind: to savour perception for its own sake.

There is a temptation for audio artists to get caught up in esoteric concerns, making those concerns the primary generator for the direction of the manifold decisions made in the construction of an audio artifact. However, "beauty is in the eye of the beholder". Audio art *is* what it sounds like to the listener, regardless of what it was meant to be. The audio artist's subjective awareness may account for many of the reasons why the artifact is what it is, but this may have little or nothing to do with the listener's relationship to the music.

While at present no responsible psychologist would claim to possess a definitive model of the perception of art, with each audio artifact an audio artist must take a position, whether overtly or by default, about the essential nature of the mechanisms for which the audio art is designed. Because the electroacoustic artist is compelled to observe neither the minimal constraints imposed by mechanical instruments nor in some respects the laws of physical acoustics experienced by composers in traditional media, constraints applied are more arbitrary and self-imposed. Consequently, a consideration of perception in relation to sonic art is all the more essential for the audio artist.

Ultimately the viability of any art form rests on an appreciation of its usefulness to its users. The essential function of any artifact of a sonic nature must be to draw listeners into active involvement with it as much as possible. Of course, anyone within range will hear the piece; there will be some low-level monitoring of the sounds and perhaps some awareness from time to time that something more or less definable as 'music' is somewhere in the environment. But that is not enough; a multitude of auditory stimuli in the daily environment register upon the mind in this way and most are simply ignored. Audio art must have the means to touch the listener's mind more deeply; to make a compelling invitation to upgrade the status of mental operations from the level of monitoring to some degree of interaction with specific features of the artifact. The greater the degree to which the listener is actively involved and attentive, the more the audio art will in some sense be appreciated.

The challenge for audio art is in finding the means to urge the listener from the initial level of simply monitoring the sound environment to the level of engaging in sufficient mental exploration of the stimuli to be able to construct auditory 'images'. In the last half of the twentieth century, the extensive presence of music in daily life has conditioned people to regard music (and audio art) in the environment largely as another type of ambient noise, and only rarely is concentrated attention brought to bear on the music – even when the potential listener is the one to turn on the radio or put a disc on the turntable.

A combination of the tendency of listeners to relegate musical stimuli to the status of background noise and the contempt with which the public may regard contemporary art create formidable obstacles to a potential listener's higher levels of processing. It is an easy matter to attract initial attention by means of sudden change, but if, in attending, the listener does not find the stimulus *worth* the attention, the result is an aggravated rather than a rapt listener. The issue is then: what constitutes 'worth' once attention is attracted? Music theorists and aestheticians for centuries have looked for the answer in the conscious mind. Whatever musical responses were, they were interpreted in predominantly verbal or symbolic terms. The problem with this interpretation is that music (and audio art) normally offers too many simultaneous layers of information and too many kinds of simultaneous processing for the conscious mind – the seat of verbal and symbolic operations – to cope with. On the other hand we can find the experience of background music beneficial without devoting conscious attention to the music at all. One cannot deny the importance of the listener's conscious attention to audio art, but the conscious processes of mind are only a small portion of the mental activity that comprises the necessary kind of listening.

Many of the requisite features of the listener discussed here are addressed in the concept of *preconscious processing*. Dixon[17] suggests that conscious perception depends upon a number of preconscious stages which go beyond simple monitoring to perform structural analysis, semantic analysis, lexical access, and emotional classification of sensory inflow; he also suggests that these stages may result in a conscious percept or that in the end they may fail to achieve conscious representation even after processing.

Audio is addressed primarily to these preconscious systems of mind. While in language the stimuli normally come from one easily identified source at any given time and the 'content' develops in essentially a linear fashion, music characteristically has *multiple* sources, each of which may be engaged in distinctive, even conflicting, lines of development. Any particular line of development may shift to another source or the role of any source may instantaneously alter. But even when there is only one apparent source, the preconscious is free to search through the many features of timbre, rhythm, tonality and loudness, and the myriad implications of the accumulation of this information over time. Because audio art allows for much more detailed and wide-ranging control of any of these features of sonic experience than does music, the audio artist's awareness of the mechanisms of the preconscious is immensely important.

Figure 1 is an approximation of the sequence in which a listener processes musical stimuli; it shows the progression from elementary preconscious processes to sophisticated interpretative operations in the conscious:

preconscious processes

1. most or all physical features are detected
2. some of 1 are evaluated for general classification
3. some of 2 are processed further for structural content
4. some of 3 are compared with similar features and judgments or predictions are made on that basis

conscious processes

5. some of 1, 2, 3, & 4 may enter to some degree into conscious consideration

Figure 1. Hypothetical sequence of auditory processing

Preconscious operations are simultaneous, multitudinous and rapid while conscious operations are linear, limited and slow. Conscious operations are those with which we try things out, speculate what might be, and what might have been. If audio art works through the preconscious mind to the conscious, it must present the kinds of structures that engage higher-level preconscious processes which in turn will pass the 'more interesting' features on. Moreover, those stimuli must have the kinds of structure that offer objects to be 'played' with at the conscious level.[18]

Given the above framework of preconscious and conscious processing of sonic stimuli, let us return to the consideration of the 'worth' that might be offered to the listener, once attracted to a work of audio art: the reason why one might *want* to 'play'. The sources of worth are many and complicated, but the key is to be found in the nature of exploration.

A framework for exploration in electroacoustic music

It would seem that exploration relies upon three frames of reference: one as a base from which to start, another for initial processing, and a third for deciding what will be processed further. In an electroacoustic work, the base might be timbral, rhythmic, tonal, dynamic – even a recurring silence. It is important that there should be a point of reference to afford some degree of security. Discovering oneself in the midst of a large forest with no idea of the way out, one is hardly likely to find the wild flowers beautiful, the bird calls fascinating, or the trees majestic. However, with the introduction of only an occasional trail marker or sign post, one can move into a completely different mode, to enjoyment of the flowers, birds, trees and other features of the forest. Because electroacoustic resources offer the potential to create particularly exotic forests of sound, there is compelling need to consider a modicum of security. It is important, however, that it be understood that a point of reference need not be familiar timbres, rhythms or tunes. The reference need only be a process that is comprehended by the listener. A person in the forest finds security in the *function* of foot paths or sign posts, not in the paths or posts themselves.

Armed as we are with the concept of exploration as a compelling and systematic activity, we are at least ready to consider how experience in audio art might be deemed to be 'useful'. Art is concerned with focussing exploratory behaviour by offering a carefully fashioned middle-ground between the meaningless and the obvious; that middle range of stimuli that demand attention and consideration. It is useful to recall Shklovsky's words: "Art exists . . . to make us feel things . . . to give a sensation of the object as seen, not as recognized . . . to make things 'unfamiliar,' . . . obscure . . . to increase the difficulty and the duration of perception . . . The act of perception in art is an end in itself and must be prolonged."[19] Art is concerned with that particular kind of exploration that systematically extends perception; a kind of exploration called 'play'. Despite the fact that we use that very word to describe the performance of music, we have long taken art too seriously and failed to appreciate how serious the business of play really is. The kind of play that art at its best offers thoroughly satisfies our craving for exploration while it efficiently exercises our powers of observation and evaluation.

An audio artifact is in some important ways analogous to a children's playground. Audio art provides resources to be used both in ways anticipated by the designer and in ways invented by those who 'use' it. A good playground provides the potential for things to *do* and so must an audio artifact if it is to function well. A playground provides the resources for a variety of activities, but the source of energy for those activities comes from the player. It is certain that the player will grow tired of any activity in time and, if the playground is to be effective, it must offer attractive alternatives in order to continue to motivate the expenditure of the necessary energy. Play is, of course, entirely possible in any kind of environment; a playground is not a necessity. However, a playground is a place that does more than allow play; it encourages three kinds of extended play:

1. *playful activity* during which a child is relaxed and happily engaged in a game involving familiar toys
2. *specific exploration* in which attention is focussed on a particular new toy and the child is actively investigating its properties
3. *diversive exploration* in which enjoyment of familiar toys is apparently temporarily satiated and the child is looking for something else (undefined) to do.[20]

A listener determines the 'usefulness' of a work of audio art on the basis of how effectively it allows him or her to engage with it in all three of these ways.

The worth of the artifact lies in the raising of many and varied expectations. To establish the base for exploration there must be something that offers familiarity in some sense, but there must also be a degree of discrepancy that encourages investigation. If a sufficient accumulation of experience triggers a willingness to explore the artifact and initial exploration suggests one or more impending possibilities, inquiries arise: "How will tensions and discrepancies be resolved?", "Where is the process going?", "When will it get there?" and "Is it indeed a simple process or are there new aspects emerging?". And when the questions are answered or when they cease to be of interest, there must be a spectrum of alternatives for the diversive exploration that will lead to further playful activity and further specific exploration.

Although a work of audio art must offer many possibilities for diversive exploration, the larger part of exploration takes place at the 'specific' level. Because audio art is something that exists in time, exploration of sound stimuli goes beyond the individual moments that pass at lightning speed. Enjoyment is to a large degree a concern for what has been experienced by the attentive listener in past moments and what may or may not be experienced in future moments. When the mind is drawn to recalling what *has* already happened, compares that to what *is* happening, in order to consider possibilities for what *will* happen, the listener is well beyond simple moment-to-moment monitoring. It is apparent that audio art is an active process of the mind, not a thing which exists as an object in a world where minds are optional.

Clichés, 'pretty' sounds, or 'aggressive' timbres and rhythms are a few of the more obvious examples of sources that might provide the basis of 'playful activity'. Familiarity in the case of the cliché and the pleasure of the sensations of the others are the features which draw the listener beyond monitoring to processing, primarily of type 2 or 3 of the five listed in Figure 1, perhaps with brief excursions into 4 or 5. This playful activity is comfortable, even pleasurable, but it is insufficiently challenging to engage attention for very long. When attention wanes, mental focus may drift from the features of the work altogether, but if the work offers an array of contrasting materials, brief periods of diversive exploration may lead to another level of playful activity. Both diversive exploration and playful activity are important aspects of the listening experience, but the listener's attention wanders rather freely and refrains from dwelling very long on any particular feature of the work.

In order to encourage greater processing of the information, the artifact must invite the listener to explore, examine, turn, and ponder individual features at some length. The particular 'worth' that the artifact can offer is in providing the basis for specific exploration. This is elicited by stimuli that have not only sufficient complexity to require the listener's sustained attention, but also sufficient promise of comprehensibility to encourage the listener to apply sustained attention. Continued attention is particularly encouraged when the music stimulates many and varied expectations. We noted earlier that the amount of prior knowledge determines how much of one's cognitive capacity may be allocated to observations. However this knowledge need not come entirely from experience prior to an individual work. If a particular sound-event is observed and pondered, it becomes potential prior knowledge which may be brought to bear at a later point in a work. While the listener's broader experience may be used to good effect, the more possibilities there are for utilizing observations made in the course of the work, the more assured opportunities there are for the listener to engage in higher-level preconscious and conscious processing.

Prior knowledge has an especially crucial function in audio art. Unlike art forms in which an observer can cast his or her eyes back to portions of the work already experienced in order to obtain further information or to confirm information already gathered, audio art offers only fleeting experiences; one cannot cast one's ears back in time. The mind engages in an attempt to understand each moment in terms of that moment's relationship to the whole work and yet the whole work is not available at any given time; only 'now' is available. As the music is experienced, the mind moves through each successive 'now', and hypothetical wholes of various sizes are based upon previous 'nows', the present 'now' and possible future 'nows'. But enormous amounts of information stream in at every moment while the mind selects which aspects to attend to. The essential nature of music is that fleeting quality of 'now'.

One of the more basic ways of maintaining mental order for these instances and making them a part of prior experience is for the listener to seek to identify cause and effect relationships. The drive to understand sonic experience in terms of cause is well illustrated by the very strong tendency on the part of listeners to try to identify specific instruments in conventional music and to imagine a great variety of fanciful sound sources in audio art, where the actual sound sources are generally outside the experience of the listener (although there are now a number of sounds which are sufficiently familiar to the general listener to be identified as made by a synthesizer). It is, in particular, in the responses of listeners to electroacoustic music that one observes this need to find some *reason* for the occurrence of particular events – a reason for those sounds having come under observation. The less such sounds are a part of a listener's experience, the greater is the need to place conscious focus upon them. Failure to comprehend or imagine causes or reasons deflects the listener's attention elsewhere. Should comprehension not be found elsewhere in the musical texture, interest in the music may altogether wane.

We observed earlier that language is largely concerned with three basic questions. Here these questions are repeated, but have been modified to apply more specifically to audio art: "What is the source of a particular [sequence of] sound?", "Why is this [sequence of] sound being made?" and "What does the [sequence of] sound mean?". Earlier, we concluded that the third question does not easily apply to audio art, but we have seen that the first plays a rather basic role, particularly with regard to using familiarity as an entry to those levels of preconscious processing above monitoring. If the third question applies to music at all, it is in the context of the second, dealing with the issue of 'why?'.

Progression and audio art

The operation of the principle of cause and effect is not limited to the question: "What made the sound?". "Why was it made?" is an even more engaging question. It is in this regard that progressive sequences play an essential role. To take an elementary example: a crescendo allows a listener to develop an awareness of the increasing likelihood that the next moment will be louder and that this increase will continue in succeeding moments moving towards some form of culmination. Knowing about this ongoing process assures the listener that he or she is following at least one aspect of the musical construction, but more importantly the process engages the listener's attention not only in the present but in the recent past, the more remote past and, to a certain extent, the future. Progression provides a means for getting beyond 'now'. Characterized by perceptible repetition of certain features while other features alter by successively greater degrees, it is a careful balance of sameness (for the sake of reference) and newness (for the sake of discrepancy), moving with sufficient consistency in directions which encourage the listener to engage in anticipation of the outcome. A variety of types of progression (particularly harmonic and melodic, but also rhythmic, timbral and textural) have worked very well to engage listeners in traditional musical idioms and there is clearly no reason why progression cannot continue to be used in audio art.

An important type of progression in audio art is transformation: the change from one distinctive state, identity, or function to another. By maintaining some features of the material while progressively altering others, the monitoring, recall and predictive activities of mind are engaged, the more so the more parameters are involved. Since the time when computer signal processing became practical, it has been possible to transform human voices to instrumental sounds and vice versa. The familiarity at either end of the transformation spectrum and the freshness and magical qualities in the middle provide a basis for a great deal of both playful activity and extended exploration.

Of great importance is the balance of simultaneous processes. Individual features or processes of a work must proceed simultaneously without degrading the integrity of one another (except where that may be a useful line of development in itself), but at the same time there must be provision for diversive exploration and choice on the part of each listener as to which features or processes will be explored at any given time. It is also important that any given process is not simply a straight line from A to B. The greatest potential *worth* of processes and transformations lies in their ability to offer at once a reasonably clear general direction while allowing the listener to move through the process in any number of ways by attending to any of the various layers of process. The audio artist's task is to provide a sufficiently open-ended structure to make the listener's contribution significant without frustrating participation; to provide sufficient guidance for the listener to find his or her way through the sonic forest and yet sufficient challenge for the forest to be enjoyed and perceived as being worth many visits.

At its most fundamental level, then, audio art is a means of satisfying a basic craving. The points we have raised regarding audio art can be applied to any category of music. If there is an important difference between electro-acoustic music and other genres, it is that the diversity and precision of the medium gives electroacoustic music vastly greater potential for achieving the objectives common to all music.

The instrumentality of electroacoustic music

The factor which is most significant in distinguishing electroacoustic music from such other music as gamelan, Dixieland, South Indian raga, madrigal or post-romantic symphonic music is the nature of the instruments employed to achieve the musical objectives of each. Although there are many cultural and stylistic differences as well as differences of purpose among these, such differences are to a significant degree formalized in the design of the instruments, and to an equally significant degree the differences are influenced by what these instruments can and cannot do. Electroacoustic music is a form of musical invention that exploits the potentials of electronic circuits and transducers to create and/or modify sound. A functional aesthetic basis for electroacoustic music may be constructed by determining how these circuits and transducers are applied, and may potentially be applied, to the creation of effective musical stimuli.

The computer is in this sense a tool. Like any other tool (or musical instrument) it is a device which extends the capability of the human body. If we look closely at the full import of what can be done with a tool, whether we are considering a hammer or a computer, we see in the tool the embodiment of the *conception* of the task; and yet the very conception of the task is both illuminated and obscured by the nature of the tool. One would not say that the computer is *merely* another tool of the composer like the piano, manuscript paper or counterpoint; a better way to express the relationship is to say that the piano, manuscript paper or counterpoint are embodiments of, and stimuli for, profound thinking – and this applies equally to the computer, the synthesizer or the tape recorder.

The nature of the next generation of electroacoustical tools and what we will do with them very much depends upon what we determine that the tools *should* do. In general, when old tools fall short of achieving new goals, new tools are fashioned, but as a matter of course they are fashioned on principles established by the experience of the construction and use of the previous tools. A culture's conception of what may be done is framed in terms of what its tools have previously permitted and the manner in which users have interacted with tools. Today's tools for electroacoustic music have come in through the side door from electrical engineering. Without even cursory consideration of the implications of such modes of thinking, waveforms and oscillators have become automatically a part of computer music systems and they may well be equally a part of the next generation of electroacoustic musical systems. During the 1980s the vastly greater part of the attention of those working in the field of electroacoustic music has been devoted to the consideration of how to achieve a physical manifestation of the gesture of the composer's hand. The perceived requirements of the tasks for which electroacoustic musical systems might be employed have been essentially limited to this function. Little attention has gone to the *purposes* which might guide the composer's hand, the *cognitive consequences* of the physical manifestations of electroacoustic musical systems, or the interrelationship between these purposes and consequences. Yet, unconsidered features of a tool are likely to be repeated over and over and unconsidered features are likely quietly to impose their consequences on the execution of the tool's tasks for many years.

At the beginning of this chapter we noted that the introduction of the tape recorder and the tone generator produced potential untempered by empirical tradition. The introduction of the computer has only *expanded* this potential, but there has been little reflection about which aspects of this potential are actually appropriate to musical experience. We have seen strategies used in the design of computer music facilities that owe more to past experience of musical notation or early analog hardware than to what could be identified as the relevant needs of the system or its operator. We seek to transcend the rigidly restraining shackles of traditional notation and the over-specialization of the oscillator, the amplifier and the filter only to build the self-same limitations into systems, ostensibly because such carry-over is necessary for novices to comprehend the new systems. We hail the new resources as a release from the bondage of the limitations of acoustic instruments only to put the larger part of our energies into developing programs which will approximate to acoustic instruments. We test our achievements not by seeing in which modes we can circumvent inconvenient mechanical constraints of acoustic instruments but rather by trying to convince ourselves that a computer-generated sound could pass for an acoustically-generated original.

A serious evaluation of the implications of the origins and development of the instruments which make up the arsenal of electroacoustic music resources raises a great many more issues than the above.[21] However, a serious evaluation of the potentials these instruments afford for more effective use of present systems and the design of future systems is even more important. Otto Laske's statement that designing a computer system for electroacoustic music is more the concern of cognitive than of acoustic engineering is certainly correct.[22] The problem is not, however, that consideration of the cognitive implications for electroacoustic music is opposed by workers in the field; it is much more serious than that. The problem is that consideration of the cognitive implications of electroacoustic music seems to be taken for granted. Until traditions *are* challenged, they will persist.

Conclusion

In conclusion we can offer no absolutes or hard facts; only impression and opinion. But then that is what the province of aesthetics is: seeking to understand the nature of our impressions and the reasons for our opinions. Perhaps this chapter has done no more than unsettle what might be taken for granted, and if so these words will have at least fulfilled Shklovsky's mandate to make things 'unfamiliar' in order to increase the duration and value of their perception.

Earlier in the chapter we were on the verge of declaring that the idea of a 'language of electroacoustic music' is an absurdity, but we found, while looking closely at language, how crucial were the words that we used *about* electroacoustic music. Later we saw that the tools selected to meet a task could actually change the nature of the task in the way that words selected to stand for something could actually change the way in which we perceived that something. The most important tool for establishing an aesthetic of electroacoustic music is *language*. We must have words to express and explain what we do as much as we must engage in the doing. Just as we must evaluate and re-evaluate the tools for making electroacoustic music, along with the tasks those tools are meant to meet, we must, in our use of language, accept neither a trivial implication nor a broad one without serious reflection of what language does to the thing so represented and vice versa.

We spoke earlier of prying open doors of knowledge, but perhaps what we have learned here is that the doors very likely already stand open. We do not need levers for prizing them open as long as we take the care to use language in such a way as not to *close* them.

6

Language and Resources: A New Paradox

Bruce Pennycook

Introduction

The elevation of orchestration from an essentially contributory role to the foreground of the compositional process predates the development of electroacoustic resources. For example, it would be very difficult to separate pitch–rhythm content from timbral content in many well-known pieces in the orchestral repertory; the exciting musical imagery in Berlioz's operatic work *Les Troyens*, the unprecedented sense of timbral motion in Debussy's *Jeux*, Stravinsky's rebalancing of the orchestral centre-of-gravity in *Le Sacre du Printemps* (and, later, *Agon*), clearly indicate that composers have explored temporal organization of instrumental colour as a primary compositional element. However, the function of timbre in these works differs significantly from its role in pieces where timbral relationships form the *principal* compositional elements. Colour melodies, or *klangfarbenmelodie*, dominate much of the electroacoustic repertory both as a procedure and as a descriptive label. Perhaps more than any other composer, Edgard Varèse, with his concepts of 'sonic structures' and his interest in sound generators, led to an intensified exploration of timbre. John Cage's fascination with non-traditional sound sources and pioneering work with radiophonics further reinforced the growing predominance of timbral manipulation in contemporary music.[1]

The development of the electronic phonograph, electronic tone generators and modifiers, and audio recording tape in the first half of this century offered composers totally new means with which to construct acoustical materials. In 1968, Pierre Boulez offered the following assessment of the potential significance of this:

> We are at the edge of an unheard-of sound-world rich in possibilities and still practically unexplored. One only begins to discern the consequences implied by the existence of such a universe. Let me note . . . the happy coincidence supervening in the evolution of musical thought: that thought is found to have need of certain means of realization exactly at the moment when electro-acoustic techniques for supplying them are available.[2]

It is at this point that I wish to begin an examination of some of the dynamics of the distribution of these new resources. I will also examine the impact that timbre-dominated composition has had on music research, and the development of commercial instruments as it relates to electroacoustic music in the educational system in America. The questions I will try to address arise more from observation – mostly in the USA – than from systematic analysis.

The paradox

An overwhelming variety of electronic, electro-magnetic, analog–digital hybrid, and all-digital devices are available at this time. These resources are rapidly being augmented by increasingly specialized microprocessor-based musical devices and computer software products. Many will become obsolete. Even more will become unfashionable. Much of the impetus for this rapidly accelerating development has come from a sustained emphasis on timbral uniqueness as the primary aesthetic factor in our assessment of new works. I have indicated in the introductory paragraphs that the forces behind this have not originated with electroacoustic music. Nor has this aesthetic been restricted to electroacoustic composition. However, the resources of electronic and computer music studios, whether adopted from other existing technologies or invented solely for musical purposes, undeniably exhibit a bias toward the invention of new sonic resources. Certainly other aspects of composition have been addressed by such developments as sequencers, random number generators, composition-algorithm software, and computer graphics programs for music notation, but none of these has received the attention focussed on the tools for sound synthesis and timbral modification.

Many composers of electroacoustic music seem to participate in a perpetual race against each other to create unique, readily identifiable 'sonic signatures'. They are also in a more threatening race. That is, the race against the commercialization and subsequent popular use of electroacoustic resources.

The *paradox* is this: electronic and computer music centres traditionally reside within universities or other government-funded institutions. Every important new technique developed by composers (a product more often of artistic than of scientific inspiration) has been absorbed by the commercial music industry. Yet it is increasingly difficult for the majority of music studios to purchase the new commercial devices. At the same time, young composers, hungry for high technology, are obliged to pursue commercial music media in order to gain access to expensive, state-of-the-art instruments and sound recording equipment.

It is true that digital equipment such as the Yamaha DX and TX series of sound synthesizers offers capability for sophisticated tone-generation at a surprisingly low price. Its inexpensiveness has resulted in unprecedented distribution of this technology to every domain of music – from rock bands to church choirs. There is no ostensible reason why composers of concert music, as opposed to pop musicians, should not use these instruments in highly creative ways, stretching the technology to its limits; they should be able to explore the gestural capabilities of the machine within musical settings entirely foreign to the pop musician. The presence, for over a hundred and fifty years, of innumerable pianofortes of all sizes and qualities has not impeded the creation of new and important works for that instrument. Similarly the omnipresence of DX-7s (over 100,000 sold), for example, might be expected to precipitate a diverse and rich repertory of FM synthesizer music.

I predict that this will not be the case. In order to support this prediction it is necessary to trace the history of the accessibility of compositional facilities for electroacoustic music and the subsequent distribution of electronic musical instruments throughout the musical community as a whole.

Composition and research

The early studios

An initial wave of technological achievements, including the theremin (1920), the ondes martenot (1928), sound-modulated light on optical film, public radio broadcasts (1926–7), the Hammond organ (1934), and, later, magnetic tape recording in the 1940s, provided the basic tools for compositional experimentation with electronic and radiophonic media. These inventions, among others, were stimulated by a world-wide enthusiasm for all kinds of electronic devices. Predictably, that enthusiasm was shared by composers who, through contact with the new technologies as radio technicians, broadcasters, or film composers, could foresee new timbral potentials.

The earliest tape pieces were constructed by Pierre Schaeffer, a broadcast engineer in the Radiodiffusion-Télévision Française in 1948. A year later he was joined by Pierre Henry and founded a studio for tape composition. In 1951 the Studio für Elektronische Musik was founded at the Westdeutscher Rundfunk in Cologne and was followed by the Studio di Fonologia Musicale at the Radio Audizioni Italiana in Milan in 1953. The Columbia–Princeton Electronic Music Center was founded in 1951 as a research facility jointly developed and shared by the two universities. In 1961 the Studio voor Elektronische Muziek was founded at the Rijksuniversiteit te Utrecht. Numerous centres for research and composition with electronic means opened during the 1960s (Paris, Stockholm, Padua, Toronto etc.), contributing to the rapid growth of electronic music composition and teaching, and to the invention of specialized equipment for both synthesis and manipulation of sound signals.

Collections of radio electronics – oscillators, filters, variable-speed tape recorders – were augmented by complex, expensive devices like the RCA synthesizer developed at the Columbia–Princeton studio, and the novel array of voltage-controlled instruments invented by physicist and musician Hugh LeCaine at the National Research Council in Ottawa. Inevitably, a widely accepted aesthetic position emerged amongst the community of electronic music composers. The 'discovery' of new sounds through the manipulation of tape, or the construction of custom devices to generate unique timbres, grew to be as important for the success of the piece as matters of temporal context. The tools of the electroacoustic music studio promised absolute control of each and every aspect of the compositional process, including minute details of the acoustical signals. It had become possible to construct entirely new worlds of sound which required little more than a tape machine, amplifier and loudspeakers for reproduction. Concerts of tape music, with somewhat perplexed audiences watching tape reels turn, were established within the curricula of music schools and conservatories.

Commercial synthesizers

With the commercial development of reasonably compact voltage-controlled synthesizers in the late 1960s,[3] it appeared to composers and music educators that a seemingly boundless new vista was unfolding. Works such as Morton Subotnick's *The Wild Bull* and, in a more popular vein, Wendy Carlos's *Switched-on Bach* (complete with glowing sleeve-notes by Canadian pianist, Glenn Gould), promised a new age. Home hobbyists could abandon their harpsichord kits and build modular, kit-form synthesizers.

It was at this point that an increased distribution of electroacoustic musical devices began to affect the role of the electronic music studio in the institutions. The Mini-Moog and other keyboard synthesizers became essential equipment for high-powered rock bands and for professional studio sidemen. Certain characteristic sounds, especially the filter envelope controls, became prominent features of bands such as Emerson, Lake and Palmer. By the early 1970s rock keyboard players such as Rick Wakeman, originally a member of the British progressive rock group Yes, had developed elaborate quasi-theatrical performances incorporating electronic keyboard instruments of various kinds stacked up like the manuals of a great cathedral organ.[4]

Electronically synthesized music had become public property. Commercial manufacturers quickly learned that most pop musicians wanted portable, easy-to-use machines that could be readily blended into an acoustical texture dominated by the electric guitar. In certain cases it became difficult to differentiate between a guitar solo and a virtuoso pitch-wheel controlled keyboard technique.

As the instrument designs began to conform more and more to live pop music performance needs, they became less and less interesting as resources for electroacoustic composition. Although their relatively low cost made them attractive, it had become very difficult to extract unique sonic materials from small commercial synthesizers which were therefore abandoned by composers. There are no pieces by Stockhausen or Berio, two of the foremost pioneers in the field, which utilize these machines. Instead, large, voltage-controlled synthesizers such as the Synthi 100, Moog IIIc and ARP 2500, consisting of a wide variety of interchangeable modules, were purchased by many universities to form the technological basis for instruction in electronic music. Students at most major music schools during the 1970s were able to experiment with the often overwhelming complexities of the synthesizer in combination with the 'classical' tape-composition techniques developed by Schaeffer and Henry twenty years earlier. Yet, fewer and fewer *major* works appeared within the tradition established by the early studios. Although an increasing number of students chose electronic music courses the musical language which they used was based on electronically generated rock or jazz styles with which they were most familiar. As studio budgets dried up in the late 1970s, the problem grew more acute. The expectation of the incoming students, coupled with a reduced interest in the hardware by composer–teachers, generated a conflict of purpose.

Widening the gap

The quest for new acoustical resources inevitably led to the theoretically unbounded capabilities of computer synthesis and computer-aided composition. There can be no doubt that part of the attraction of computer music has been that, until very recently, this technology was available only to a small sector of the musical community. Sound synthesis languages such as MUSIC V, written by Max Mathews at Bell Laboratories, and Barry Vercoe's Music 360, an IBM adaptation of MUSIC V concepts, required access to large mainframe computers which could accommodate digital-to-analog conversion hardware for real-time playback. Institutions unable to gain access to such equipment were forced along one of three paths: continuing analog electronic music courses as long as the equipment survived, abandoning the area entirely, or developing more modest hybrid systems in which smaller computers were used to provide control signals for analog synthesizers.

Other factors have contributed to the widening of the technological gap. Most synthesizers can be used as the basis for trial and error. Even a simple patch can instantly yield rewarding results. Knobs and other controls are designed to suggest their function in visual and tactile ways. Computer music technology, on the other hand, poses formidable obstacles. The computer music novice has to assimilate an enormous amount of information and invest many hours before even the most rudimentary sound can be coaxed out of the machine.

Computer music composition and research has, by necessity, been centred at institutions which already had substantial computing facilities. Even though at least one of the various sound synthesis languages could be run on the large mainframe computers which most universities maintain, the specialized demands of real-time audio data conversion posed technical problems far beyond the capabilities of most music departments. In many cases, conversion of the digital sound samples written to magnetic tapes by the mainframe system, has required access to yet another computer facility which could provide these services. From the time the data for the compositional and synthesis aspects of the program were submitted to the mainframe to the actual moment of audition, days or even weeks may have passed. (Imagine walking up to your piano blindfolded, playing a few minutes of your new work and hearing it days later in a different room!) The patience and tenacity of computer music composers working under these conditions is to be lauded. Nevertheless, there are precious few minutes of music for all the accumulated hours of work. It is not surprising that only those composers and students with both the aptitude and desire were able to endure these circumstances.

The capital investment and personal commitment required to build a computer music studio created a situation similar to that of the early period of tape music studios: only those institutions with sufficient financial and human resources have been able to secure or share the necessary equipment. The exclusivity of the field, however, virtually guaranteed that those with access to the technology would enjoy a high professional profile. Just as electronic music spawned a society of 'techno-haves' in the 1960s, computer music has generated an even more remote community.

The distribution of computer music technology

Research in computer music did not ultimately flourish on the large mainframe systems except in a few major centres. It was necessary for composers to align themselves with either Computer Science or Electrical Engineering departments (or both) in the early years of the field in order to gain access to more flexible technologies. The Digital Equipment Corporation PDP-11 series of computers, the favourite mini-computer of the 1970s, was well-suited to real-time operations. Its low cost in comparison with an IBM 360 for example, combined with more accessible operating system designs, presented more suitable environments for digital audio experimentation. Yet with the exception of MUSIC 11, written exclusively for the PDP-11 by Barry Vercoe at the MIT Experimental Music Studio, very little music software has emerged. The Center for Computer Research in Music and Acoustics (CCRMA) at Stanford University, under the directorship of John Chowning, also developed specialized music software. All of this work, however, was done on a larger, far more expensive DEC system: the PDP-10. Distribution of CCRMA software was limited by the necessity to run the Stanford operating system, SAIL, in order to implement any of the computer music programs. Few institutions with PDP-10s in general use by the university community were prepared to support computer music at these costs. A notable exception is Colgate University in Hamilton, New York. There, using Stanford software, Dexter Morrill assembled one of the most important computer music studios in terms of musical output in a small undergraduate college.

Computer music became firmly entrenched in those institutions which were prepared to make heavy investments in the field. Access to this technology was provided through summer computer music courses and visiting appointments. Many teachers of composition made costly pilgrimages to Stanford-CCRMA or MIT-EMS in order to attend lectures and to experiment with the acoustical resources. For many this proved to be very frustrating, as their hard work could not be transferred to their own institutions for further development. Those who elected to build computer music studios of their own soon realized that most computer centres were reluctant to install music software and audio conversion equipment which could potentially absorb most of the available resources. The alternative was to raise capital for systems exclusively dedicated to computer music activities. The start-up cost for a modest system, based on a PDP-11 for example, exceeded the budgets of all but the largest schools. Certain institutions had long histories of significant research and composition in analog electroacoustic music and, based on that, were able to raise substantial grants for computer equipment and the hiring of technical personnel. One of the most important of these in terms of overall contributions to the field is the Computer Audio Research Laboratories (CARL), at the Center for Music Experiment, University of California at San Diego (UCSD). CARL, under the direction of F. Richard Moore (a major contributor to the development of MUSIC V at Bell Laboratories and a PhD graduate from Stanford in electrical engineering), distributes a complete,

UNIX-based, computer music software environment for the VAX series of computers, which represents many years of software development.

Digital synthesis systems

Several other prestigious music faculties have installed software-based computer music studios: Northwestern University, the Eastman School of Music, Oberlin College, Princeton University, and in Europe the universities at Marseilles, Padua and Utrecht. However, most music colleges have entered the digital music domain by purchasing special-purpose hardware.

The two most widely distributed instruments in this category are New England Digital's Synclavier, developed in part by Jon Appleton, Director of the Dartmouth College Electronic Music Studios, and the Fairlight Computer Music Instrument, from Australia. Both of these systems offer impressive digital synthesis power together with digital recording, modification, and playback of recorded sounds. (The latter feature, now commonly referred to as 'digital sampling', was added later to the Synclavier to compete with the world-wide interest in the sampling capabilities of the Fairlight.) These instruments presented complete computer music environments packaged in such a way that musicians familiar with conventional electronic music devices could learn to use them quickly. They operated in real-time, which made live computer music performance practical and permitted a limited form of *musique concrète* through digital edition and processing of live sounds.

Other digital instruments appeared on the market around 1980. Digital Music Systems' DMX-1000, a kind of real-time MUSIC 11 machine designed by Dean Walraff, offers complete programmability – a highly desirable feature which most digital instruments lack. However, it requires a PDP-11 computer to serve as a host and has not therefore enjoyed wide distribution. The Emulator and Emulator II offer keyboard control of sound samples like the Fairlight but at a much lower price. The PPG Waveterm, a German product, provides performance features similar to the Fairlight plus some unique extensions. The Synergy, a compact version by Crumar of its own large General Development System, first appeared in 1981 providing high-quality FM synthesis at prices much lower than the Synclavier.

With the exception of the DMX-1000, these and other commercially manufactured digital devices provided computer music technology without the difficulties of installing and maintaining general-purpose computer systems. Not only were they attractive to music schools, they were embraced almost immediately by commercial musicians. Entire film-score or advertisement soundtracks could be assembled by one person using a Fairlight or Synclavier. Very quickly, the unique timbral capabilities of these devices could be recognized in rock recordings and television jingles. Once again, the private acoustical domain of the music institutions, which had invested vast amounts of time and money, had become public property.

Commercial recording technology

Another factor contributing to the widening gap between commercial applications of electroacoustic music technology and the presence of electroacoustic music in the schools over the past decade has been the staggering improvement in analog recording equipment. The current standard in professional recording studios includes 24-track tape machines, 16- to 32-input mixing consoles, a wide variety of precision microphones, plus numerous analog and digital sound processing devices. Successful rock bands now tour with studio quality mixing boards and microphones, together with powerful sound reinforcement amplifiers and speaker arrays. The investment in audio equipment for a modern studio or professional rock band is enormous. Added to this are the costs of pianos, synthesizers (including expensive digital instruments like the Fairlight CMI), and the host of support equipment needed for patching, dubbing, system maintenance, lighting systems, and so on. Predictably, most electronic music studios have been left far behind.

Much of the latest studio technology is digital. As early as 1979 *Studio Sound* published an interesting survey of the various makes and models of digital sound processing units available for recording purposes.[5] The article also included a comprehensive chart of the distribution of digital equipment throughout the major studios in Europe and North America. Although this list is now out of date, it serves to indicate that for nearly a decade now, digital technology has been replacing conventional analog equipment in the domain of commercial recording. This is also true for live-performance hardware. All major rock concerts use signal processing such as echo, phasing, flanging or reverberation. (It is perhaps ironic that one of the roles of the digital delay unit is to simulate the exaggerated and rather crude echoes of rock-a-billy recordings of the 1950s.) A more recent article in *Studio Sound*, 'Effects, Reverb and Equalizers',[6] presents a more up-to-date view of the rapidly expanding scope of digital (and in a few cases, analog) sound processing devices.

A parallel growth in the development of low-cost compact disc players has encouraged recording studios to invest in digital mastering recorders based on pulse-code-modulation technology. Current consumer versions employ standard video-cassette machines as the tape storage medium. These systems provide several hours of full-fidelity, noise-free digital recording at costs below high-quality conventional tape decks.

It seems unreasonable that, with few exceptions, composers and performers at most music schools cannot benefit from the impressive improvements in audio engineering enjoyed by commercial musicians. It could be argued that those concert musicians who have earned sufficient status *do* utilize the technology through recording contracts or live broadcasts, but they represent a small proportion of the recording business, and, of that small fraction, the amount of contemporary music recorded – with or without electroacoustic forces – represents an even smaller fraction.

From hopeful rock and country bands at their first recording session to the producers of slick radio and television advertisements, state-of-the-art technology is readily available and fully exploited. In fact, the quality and quantity of new equipment forms a major aspect of the recording studios' marketing strategy. It is assumed by novice and seasoned studio artists alike that the preparation of competitive products requires the latest technology. Hence, the average budget for a rock album produced in a major studio can soar to $100,000 and more. Combined with the market demand that new releases be accompanied by exotic videos, the prospects for professional recordings of musical styles outside the commercial mainstream grow more and more remote. Only the very successful or independently wealthy concert artist can expect freely to engage this technology for creative purposes.

Recently, private studios have begun to flourish in response to the limited budget of the newcomer. Tapes of very high quality (though perhaps not quite up to the quality required for broadcasting) can be made in these small studios for a fraction of the cost charged by full-size professional studios. Some manufacturers offer 2-, 4-, 8- and 16-track capability on quarter- and half-inch tape formats which helps reduce costs even further. Fitted out with digital synthesizers, digital sampling units, and digital drum machines these small studios can produce elaborate finished products with only one or two musicians.

Although these low-budget studios greatly benefit the cost-conscious musician or producer they have come into conflict with other market forces. Radio and television stations in North America are bound by contractual agreement to play only union-made recordings. Furthermore, there remains some controversy over the use of sampling machines. Like the once-outlawed Mellotron (a keyboard-activated multi-tape player), sampling machines permit the recording and playback of several seconds or several minutes of another person's musical performance. Plagiarism laws dictate that the maximum number of recognizable bars a composer can 'borrow' is eight, although convincing arguments against even a few notes could probably be made. The latest sampling equipment offers far more than single, isolated musical tones in playback mode. Nevertheless, the high cost of professional studio rental, alongside union-controlled sideman fees, will force an increased reliance on all-in-one recording studios and sampling devices.

There exists a general anxiety within the music departments about the issue of borrowing both from actual pieces and, more generally, from recognizable natural or artificial sounds. The digital 'player-piano' concept runs contrary to the 'quest-for-tones' aesthetic. The following is a statement on this topic by Boulez from *Notes of an Apprenticeship*:

> In electronic music, we must 'constitute' each sound that we mean to use. This is not a question of copying natural sounds – some of them being excessively complex, that synthesis is in practice impossible; furthermore, the very idea of creating an '*ersatz*' natural sound universe is dubious: it will have neither the quality of the original nor the specific characteristics that one has the right to expect from it. This is the sandbank upon which nearly all the specialists have run aground: hybrid, asexual, the electronic instruments have not been capable of posing the real problem of the timbre in this 'artificial' realm.[7]

It appears that the sands have shifted somewhat more rapidly than expected.

Music schools, students, and pop technology

Initially, electronic music courses were chosen by young composers eager to emulate the new music of Berio, Stockhausen, or Davidovsky. Access was limited to those who had achieved a certain standing in theory, counterpoint and orchestration, and who had exhibited advanced compositional skills with traditional musical forces. By the early 1970s, inexpensive synthesizers made it possible to offer a greater number of courses to a larger number of students. It was presumed by many educators that electronic music would flourish and that synthesizers would become as prevalent as pianos.

Queen's University in Canada introduced a *mandatory* course for all Bachelor of Music students, in addition to optional courses for interested composers and students from other disciplines. Over the period 1980–85, however, a curious shift in the population in the electronic music course occurred. The mandatory course was dropped, and it became clear that very few music majors showed even a passing interest in the electronic music studio other than for recording their recitals. The majority of students came from outside the music department, primarily from Electrical Engineering, but also from Film Studies, Drama, and other arts areas. This trend indicates that the listening habits of those students differed significantly from those of the music students.

While the film students and engineers are accumulating vast collections of rock music, much of it synthesizer-based, music students are absorbed with the rigours of Western music. Neither group absorbs a steady diet of contemporary concert music from the media or their own musical activities. Furthermore, like the rest of the listening public, neither group receives the necessary preparation to enjoy, or at least appreciate, the electroacoustic music of Stockhausen, Berio, Subotnick etc. Their musical experience (and by that I mean direct contact as selective listeners), often shares little or no common ground with the musical styles presented as exemplars of the electroacoustic genre.

On one hand, music students resist electronic music because it operates in a world based on technology they rarely use or require in their daily activities. On the other hand, students who listen to electronically produced music on a regular basis are frequently denied the freedom to experiment with forms of musical expression most familiar to them.

How can either group be properly served by electronic music as it is currently presented? Should substantial amounts of money be directed toward a receding technological target or should those schools which cannot afford repeatedly to purchase new equipment simply drop the whole issue? Shrinking budgets will force the latter in some cases. A few schools will continue to attract research funding. For the rest however, what is required is a total reassessment of the role of musical technology within the entire music curriculum.

Computers and music instruction – a question of access

Fortunately it is no longer necessary for music schools to invest in large, sophisticated computer systems in order to benefit from current developments in music technology. Recently, a number of musical instrument manufacturers have agreed upon a common mechanism for connecting together synthesizers, drum machines, sound processors, and microcomputers. MIDI (Musical Instrument Digital Interface) may significantly alter the role of electronic musical instruments in all levels of the education system and in the commercial domain. Unlike previous developments in electroacoustic music which have emerged from university research environments, MIDI has come to the institutions from industry. In fact, the astonishing growth in the production of MIDI software and hardware by both large and small firms triggered a reversal of the flow.

The following is a list of some of the problems encountered in computer music, many of which have constituted university research projects in computer music for several years. Of course, the authors of the MIDI-based solutions have borrowed heavily from previous groundwork and there remain many more problems to be solved. What is new here is that instead of isolated, unportable software, MIDI products are essentially fully compatible.

Recording musical gestures. MIDI permits convenient interconnection of micro-computers and up to 16 independent musical instruments. Technically, it is a highly specialized form of computer network.[8] Information on musical gesture, such as key depression and release timings, the velocity at which keys are struck, the setting of buttons to select voices or other timbral features, the motion of controllers such as pitch-wheels or volume pedals, or any other aspect of performance, is encoded as a sequence of digital values which can be transmitted to a computer equipped with a MIDI interface. Individual tracks, like the individual channels of a tape recorder, can be independently synchronized, recorded, and re-played. The most interesting aspect of this capability is that a wide variety of devices can now be considered 'performable' instruments, including signal modifiers, electronic drum surfaces, or even light beams bouncing off a dancer!

Music printing from performance. Many of the MIDI software packages can translate performed musical events into musical notation. Although there are still some problems with resolving rhythmic irregularities inherent in all forms of performance, careful use of these programs can yield good quality printed music directly from MIDI data. This form of 'performance-to-notation' is a major achievement which could radically alter musical instruction methods.

Automatic pitch detection. Perhaps one of the most appealing developments is the ability to translate a voice or single-note instrument into MIDI data. Coupled with a MIDI recorder or printing system, these devices automate the often tedious task of transcribing improvisations by hand. The MIDI-controlled digital sampling unit enables the nuance of a performance to be re-created with different sound sources.

Automatic accompaniment. Recent work by Roger Dannenberg at Carnegie-Melon University has produced a system which can track a live performer and modify a stored accompaniment in response to the soloist.[9] This work addresses one of the basic limitations of tape music: the accompaniment, fixed as a pre-recorded signal, cannot make allowances for the natural deviations which all performers have the right to make.

Composition environments. Multiple channels of MIDI can be used to build very elaborate compositional textures. Though direct real-time control of timbral parameters is difficult with the current generation of MIDI synthesizers, the gathering and manipulating of large volumes of gesture data (pitch, rhythm, amplitude, vibrato etc.) is relatively easy. Data can be entered from touch-sensitive keyboards, from pitch detectors, typed directly at the computer terminal and viewed as common musical notation, or generated by compositional software. Many of the MIDI-capable computers include extensive editing facilities. These enable the composer conveniently to experiment with the work in progress, hearing the revisions immediately. Even if the final version of the work is for acoustical rather than electronic forces, such a highly malleable compositional 'scratch-pad' can be invaluable. Furthermore, the signal degradations associated with repeated tape dubbing no longer occur; each new version of the MIDI-encoded composition produces first-generation audio quality.

It should be clear from this short list that the focus of computer music systems has shifted from primarily timbral considerations to gestural and compositional matters. This shift has also been accompanied by both a decrease in entry-level cost and substantial improvements to user-interface design. Many commercially available computers are either MIDI-equipped, or are readily modified by the addition of a memory card or external interface. A wide variety of MIDI-equipped synthesizers, sound processing units, keyboards and many other devices yet to be released (MIDI-controlled audio mixing consoles, and multiple keyboard stations for classrooms linked to a master console through MIDI) will encourage a proliferation of specialized, user-friendly software.[10]

It may be unwise to offer any promises that MIDI will revolutionize our concepts of music training and musical practice as a hobby or vocation. One thing is certain though: MIDI-equipped home computers will provide new ways for amateur musicians to experiment with musical *ideas* as well as timbres; it will become more interesting to compose (or at least restructure) musical passages than simply to search for 'interesting sounds'. Perhaps the whole idea of hunting for contextually isolated sonic entities will be displaced by serious study and participation in the processes of composition.

Computers and music research – What is left?

There are many problems yet to be solved. Solutions to most of these far exceed the capacities of the MIDI network as it is currently defined. As the large instrument manufacturers can be expected to continue to produce low-cost sound generators, synthesis techniques will quickly move from the research labs into the public domain. In fact, many intriguing algorithms exist which are not available in commercial form.

I expect that most computer-based research will gravitate toward the extremities of musical activity, to digital audio engineering and musical cognition. Some problems, like unravelling the pitch–rhythm content from many simultaneous instruments or voices – perhaps one of the most difficult tasks – require highly specialized digital signal processing algorithms as well as certain forms of 'intelligent' data extraction. This level of research is far beyond the training of most professional composers. It demands specialized skills in the mathematics of signal processing as well as expertise in cognitive science and artificial intelligence programming. However, the composer–theorist will play a vital role in the development of analytical and computer-aided compositional software. Few engineers or computer scientists have the depth of musical knowledge required to make significant contributions to our understanding of musical processes.

The large, well-funded labs at Stanford, UCSD, MIT, and IRCAM, amongst others, will continue to draw together specialized teams of musicians, engineers, computer scientists and psychologists to tackle such problems. In order for the smaller centres both to benefit from and to contribute to this work, they must take a hard look at their commitment to computing facilities, future hiring policies and to the presence of music technology within their teaching programmes.

If a college or other institution intends to invest in future developments in music and technology, it must accept at the outset that the technology will change very rapidly over the next twenty years. One of the reasons why so many studios are in such bad shape is that large, one-time grants were secured and spent in full, leaving inadequate funds for up-grading the technology and for technical assistance. This has never been the policy in computing departments where it is assumed that equipment will turn over at least every five years. It is simply not possible to maintain credibility without evidence that the researchers have access to the same quality of tools made available to industry, or in the case of electroacoustic music, the commercial recording studios. Yet musicians are caught between having to sacrifice precious composition time in order to develop skills in areas which receive a steady flow of government research funds, and having to convince the various institutions that their *creative* activities alone deserve substantial equipment and support-staff budgets.

One of the reasons that the sciences are well supported is that there exists the potential for substantial revenues from royalties or licensing of important discoveries. There have been few discoveries in electronic and computer music that have generated large sums, the most famous one being the Chowning implementation of Frequency Modulation which Yamaha procured from Stanford. But the commercial music industry is very healthy. Composers of film scores, jingles, and pop songs can make enormous sums. The big recording studios may not enjoy large profits, but certainly have strong cash-flow. If the invention of a low-calorie sugar or a new computer language, for example, is worthy of state-of-the-art physical plants and technical support, why then has commercial music-making, the greatest potential source of revenues for musicians and music schools, been unilaterally shunned? I realize that I am out on a limb here, but consider for a moment the history of jazz education in schools and universities.

Jazz has been legitimized by the media, by educators, and through praise from concerts musicians. Works such as Stravinsky's *Rag-time* and *Ebony Concerto* reinforced a suspicion that somehow the music of black Americans might be important. Jazz schools, jazz studies programmes, and PhDs in jazz history are now commonplace. Yet most of these were put in place *after* jazz was widely popular and commercially successful. Many colleges in the United

States hired big-name performers after their careers peaked, hoping perhaps to preserve what had become for the rest of the world 'American music'. Yet rock and its derivatives, now over 35 years old, remain forbidden.

The rock music industry, including electronic instrument manufacturing, is central to the financial growth of the music business. Classical music recordings are financed with profits from this music. Similarly, performance and mechanical royalties paid to concert musicians and contemporary composers by the various societies are heavily subsidized by these same profits. Imagine a similar mechanism at work in the music schools. Funds could be diverted from successful endeavours in the popular field to build and maintain recording studios, computer music research, library acquisitions, staff appointments, and all of the other activities which today are sorely underfunded. Graduates, fully trained in popular music activities – performance, arranging, composing, studio production and so on – could channel profits back to their old colleges and universities. Contract work and royalty benefits from the activities of full-time members of academic staff could also generate 'research' revenues. An aggressive, competitive approach in many spheres of commercial music activity could ultimately generate a significant proportion of the operating costs of an entire department. Shrinking enrolment problems could also be reversed by high-profile successes in popular fields.

This fanciful notion is predicated on much more than money. The presence of music instruction in all levels of music education is being seriously challenged by vocationally minded administrators and parents. By responding to the cultural realities of the role of music in our society, and adopting more popular components into our music programmes, we may, all at once, narrow the gap between commercial and concert music, improve the viability of music instruction (at least as it is *perceived* by non-musicians) and increase the capital base on which the system is predicated. I am fully confident that under such a system, the dedicated concert performers and composers would continue to surpass their peers and reach for more demanding musical experiences. The curious thing is that those who may agree will concede that the move is long overdue. Those that cannot accept such a premise never will.

Closing comments

The forces which have been affecting the role of electroacoustic music in our educational institutions, concert halls, and broadcasting media are a function of the cultural dichotomy discussed above. Popular music is, in many ways, a *direct product* of modern technology. The small amount of electroacoustic music on the concert stage and in the classroom has resulted from composers of concert music rejecting the sonic resources of the popular domain. While the market for digital computer-controlled synthesis equipment is exploding, fewer and fewer concert pieces emerge.

In order for any major change to occur, the resources of electroacoustic music (which at this point in history include computer equipment), must become part of our collective musical experience *beyond* popular music. Too many aspects of the basic musical skills being taught to students are being tackled by the commercial industry for them to be ignored. Students (and most lecturers) are not only essentially uninformed, but also being rapidly left behind by commercial recording artists.

I recently overheard a music professor who had been trained in Europe state that the problem with American music programmes stemmed entirely from a lack of 'musical tradition'. Although he was correct, he assumed that what is needed is for North America simply to transplant European musical tradition. Despite numerous attempts over the past century and a half to achieve that goal, it is now time to accept the fact that new traditions are emerging. Token representations of the world of electroacoustic music seem just as futile as token representations of dated jazz styles.

The *paradox* remains: with the exception of a small number of highly motivated composers and theorists who have accepted the challenge, teachers and students within the music institutions continue to reject the technology of electroacoustic music and computer music research. The students most interested in the field are denied access or are diverted from expressing themselves in the musical styles from which their interests first emerged.

This decade and the next will bring about a staggering increase in the presence and integration of powerful micro-computers, sophisticated software and specialized peripherals in all areas of human endeavour in industrialized society. I suspect that it may be naive even to raise this warning. However, we have a choice: we can ignore the flood or we can begin a systematic investment toward providing a more realistic, culturally relevant, approach to the presence, appreciation and training of electroacoustic music in all levels of our education system.

The Influence of Computer Technology

7
Computer Music: Some Aesthetic Considerations

Michael McNabb

On initial reflection, I realized that I was not sure what the word 'aesthetics' actually meant. After looking it up in the dictionary[1], I am even less sure. I find two definitions:

1. A branch of philosophy dealing with the nature of the beautiful and with judgments concerning beauty.
2. The description and explanation of artistic phenomena and aesthetic experience by means of other sciences.

Noting the self-reference in the second definition, I look up the word 'aesthetic':

> Relating to or dealing with aesthetics or the beautiful.

Aesthetics is therefore defined only as a description of itself, at the same time as having something vaguely to do with the equally vague concept of beauty. I begin to feel myself falling into the vast chasm of semantics and philosophical nonsense. Clinging to the word 'experience', I struggle to pull myself up on to the rocky ledge of meaning. The corollary to the old maxim "Beauty is in the eye of the beholder" is that "There is no beauty without the beholder (and his eye)". This seems firmly to link the existence of beauty with the perceptions, and by implication, with the resulting experiences of the viewer (or in our case, the listener). Thus, at least one measure of beauty, or of a successful 'artistic phenomenon', is how well the phenomenon is perceived, and how strongly the perceiver is affected in some way by it. This is a field in which computer musicians love to frolic. However, I would go even further. As far as I am concerned there is not only no beauty, but no art at all other than in instilling an experience of some kind in one's audience. The degree to which one understands, controls, and uses aspects of one's chosen medium to this end is the measure of one's power as an artist. Prerequisites to the successful use of this power include a consideration of the audience's previous experiences, an optimization of the context for the work's presentation, and a detached examination and understanding by the artist of his or her own internal vision. And while many of these last items may seem self-evident, one must bear in mind that they are all links in a chain of decisions and actions which lead up to the artistic experience.

Although the medium is still quite new, there are already a multiplicity of approaches to and uses of computers in music, but a rather poor understanding of what it all means for music in general. There is in fact a lot of confusion, especially following the somewhat unforeseen commercialization of certain aspects of the field. For example, whereas this development has allowed more people access to some (rather restricted) areas of computer music, through lower costs, it has in some cases distracted composers and researchers from working in what might ultimately be more rewarding areas of the field.

For the purposes of this chapter, I see 'aesthetics' to be the examination of some of the techniques, approaches, and other issues which affect either the electronic composer's expressive ability or the audience's ability to experience something from that expression. The perception of music and the resulting reaction is by definition dependent on the individual listener's associations, past experiences, and imagination. It is not therefore meaningful to talk about some kind of *general* aesthetic. In other words, any abstract music, especially in today's world, is bound to have a multiplicity of meanings for different listeners. This is what Stravinsky meant when he said that music was powerless to communicate anything. Music for me is more of a catalyst, an expression of pure imagination which somehow triggers a similar state to some degree in the listener, who has a need for it. Instead of a philosophical discussion, it is preferable to refer someone to the music itself, and let each person have his or her own unique experience. This after all is the main purpose of the whole exercise, not to provide something for critics and academics to write about.

As communication occurs more rapidly the smallest innovation turns into a tradition overnight. No one can use it or even refer to it without critics and the reigning musical establishment condemning or praising the composer for appearing either to follow or to react against it and all the other musical 'traditions'. It is an alarming and paralyzing situation for young composers, who start out innocently expecting simply to express themselves, have some fun, and perhaps add a little beauty to the world.

I believe that composers who are serious about their art must completely disregard all such nonsense. In the effort to create an 'aesthetic experience' founded on one's personal vision, it is ludicrous to base a decision about whether to use a certain style or technique on anything other than its appropriateness to the task at hand, and its relationship to the work as a whole. Truly new and powerful musical ideas can only come from the complete freedom to express oneself, without concern over whether or not one's work conforms to or avoids anything whatsoever. No one is capable of deciding in advance to produce a revolutionary work of art. It cannot be done merely by trying to devise a new sound, or a technique which opposes current fashion. History shows that the true artistic innovations were produced by those with compelling ideas, who drew freely and capably on whatever means were at their disposal to express themselves.

* * *

The phrase "The computer is just another musical instrument" is usually intended as a kind of apologia for the introduction of such advanced technology into the field. This statement is not the whole truth, however. It *is* true that the most frequent use of a computer in music today is as a sound-generating device, so this statement is useful in calming people who fear that computers can make music out of thin air, rendering humans obsolete. (The statement is frequently followed by a further apology – "and we mainly use it to do things that traditional instruments cannot do" – which is useful in trying to pacify some critics.) What usually goes unstated, though, is that for many composers the computer may be the only musical instrument available which is capable of providing them with a means of expression appropriate to their needs. For example, if your style runs toward large orchestral textures, you can write an orchestral piece, perhaps be lucky enough to receive one or two under-rehearsed (and possibly uncaring) performances, and have a microscopic chance of a recording. However, if the same piece is produced with a computer, the composer can have absolute control over the performance, and can look forward to a much larger audience, since the costs of performance and recording are greatly reduced.

There is no question that this is one of the main reasons why many composers first become involved in the field. In fact, it seems ridiculous to me that we should feel the need to make any apology at all for this use of the computer. If I want to make computer music that sounds just like acoustic instrumental music, why shouldn't I? After all, those are nice sounds, I can play the computer better than I can play all of those instruments, and I can usually get a better performance, at a great many more concerts!

This of course brings up the old complaint: "But isn't there something missing when there's no performer up there on stage?". I consider this to be a misplaced criticism. Every day untold numbers of people are greatly moved by music which they have never heard performed live, via their home audio system. They rarely exclaim afterwards: "I would have enjoyed that so much more if the entire Chicago Symphony Orchestra had been here in my living room." When we view a painting, we do not insist that the painter be there slapping paint onto canvas. And how many times have you gone to an instrumental concert and kept your eyes closed a lot of the time? Frank Zappa has actually demonstrated that musicians miming a recording of synthesized music can completely fool an audience and professional critics. Personally, if I want to watch something, I prefer to go to a movie.

The reason that a lot of tape music sounds unsatisfactory is not because there is no performer on stage, but simply because there is no performer at all. At an orchestral concert it is not the presence of live instrumentalists that matters as much as the fact that there is a performance, with all that that implies in the way of the added subtleties and dynamics of expression in the music. There is no reason why there cannot be a musical art which is the end result of a kind of sculptural process, but composers of electronic music must realize that they are the performers, and are therefore responsible for adding all the nuance of performance to the music if there is not going to be someone at the concert to do it for them. The composition process must extend down to subtler levels. Otherwise the end result is a kind of 'audible score', instead of music. Of course, this implies a lot more work, but that is the nature of the medium at this time.

Another source of this criticism has to do with the circumstances of the performance. Electronic music must be presented with every bit as much care and rehearsal as instrumental music performances. An inadequately powerful and unclear sound system, or one unable to adjust to the hall, or insufficient time for adjustment, can effectively sabotage a piece of tape music. Organizations which attempt to put on concerts without due regard to these issues are irresponsible and often do more harm than good. Even well-known institutions with a reputation for producing good electronic music sometimes have shockingly bad performance practice. Educational institutions should provide more training in this area, in order to raise the general standards of electronic music performance practice and allow composers to breathe a little easier when they send tapes off for a performance.

Another counter-productive practice that deserves a merciful end is the 'informal' tape concert, with people coming and going, scheduled at odd hours, or even at the same time as other 'formal' (i.e. instrumental) works. This kind of event doesn't give the music a chance to be heard properly, and only serves to reinforce the misconception that electronic music is somehow inferior. Unless it was specifically composed for such a situation, tape music should always be presented with the same respect and care due to the performance of any other music. Composers should be wary, and always insist on the highest possible performance quality.

It is a welcome development that consumer audio equipment has reached very high standards of quality in recent years. With the introduction of the digital audio disc, electronic composers can avoid the trials and tribulations of public performances, and deal more directly with their audience.

Of course, the computer as instrument really can do things that conventional instruments cannot do, and this is precisely one of its strongest sources of musical power. However, to me the power lies not simply in the availability of new sounds, but in the careful and precise use of these sounds in the context of familiar musical elements.

All perception requires a frame of reference to be meaningful. Our minds constantly seek relationships in everything we see and hear, comparing and contrasting our current experience with previous experiences in order to increase our understanding of both. Because of this natural tendency, art which suggests a relationship between otherwise unrelated elements can be very compelling. For me, it is especially so when one of the elements is rich with associations for the viewer or listener, and the other is new, or strange, or somehow mysterious.

In computer music, one can take advantage of the unique plasticity of the medium to produce combinations of and transformations between arbitrarily different sounds. Again, it is the interplay of the familiar and the unfamiliar or unexpected which I find to have the most expressive potential. A juxtaposition of two completely unlike sounds need not be instantaneous, but can occur over any time-span, producing a gesture of great musical expressiveness and beauty, poignancy or tension, a concept which was the primary source of inspiration for my work *Dreamsong*.[2] This juxtaposition of the familiar and the unfamiliar seems to me to be of great historical significance. Greater, say, than the introduction of the crescendo to Western orchestral music.

Context is very important to the success of such a gesture. A radical transformation is at once both a compelling and a subtle event. It is not usually something whose progress or conclusion can be anticipated by the listener (as can be done with tonal harmony, for example), so it requires complete attention. There should not be very much else going on musically, since at one extreme it may go unnoticed or at the other be too distracting. For example, one of the most effective aspects of a straight transformation between two sounds is that there may be no identifiable point where it stops being the first sound and begins to be the second. If the listener's attention is distracted from the transformation for even a brief moment, the effect can be lost.

There are many more subtle applications of these concepts. One of my common practices is the use of dynamic combinations of timbres based on conventional acoustic instruments and voices. I have a synthesis algorithm which allows me to specify several basic timbres, whether based on natural sounds or invented, and a function which can specify any interpolation path between any number of mixes of the timbres. The interpolations from one mix of timbres to another can occur over one note, over a phrase, within one voice or among several. Alternatively, the interpolation can be controlled by some kind of time-varying function, a process which I call 'timbre vibrato' due to the natural analogy with frequency and amplitude vibrato.

Another variation of the use of timbre is the breaking up of the spectrum of a sound and the use of its elements in such a way as to blur the distinction between timbre and melody, or timbre and harmony. For example, a wonderful and expressive effect occurs when the harmonics of a synthetic instrument are fragmented so that the overtones go on and off rapidly as if they were notes in themselves, at the same time as the whole overtone structure follows the frequency shape of a more slowly moving melody. Further, since computers allow the arbitrary and instantaneous definition of tuning systems and harmonic structures, entirely new musical structures are possible which have the beauty and clarity of just intonation, but which, like John Chowning's *Stria*, need not be based on the harmonic series. Even the manipulation of one single timbre can effectively serve as the basis for an entire work, as in Chris Chafe's *Solera*.

Any musician will agree that it is the subtleties of pitch and amplitude variation which give a musical note or phrase its expressiveness. The better performers also know that fine control over the timbre of their instrument can add greatly to that expressiveness. The computer is the first instrument to allow as great a control over timbre as it does over the other elements of musical sound like pitch and amplitude; this on its own increases the range of musical expression. The use of non-pitched sounds and their arbitrary and seamless blending with pitched sounds increases the range again. When you consider that 'non-pitched sounds' includes everything from a few milliseconds of noise to minutes of recorded environmental sound, the possibilities seem vast. If a blank sheet of music paper was the most frightening object for Stravinsky, a blank computer music system screen would have been absolutely terrifying.

Much used to be made of the control over apparent spatial location which computer music provides, especially using four loudspeakers. Early pieces by John Chowning and others demonstrated that under carefully controlled conditions an effective illusion of a moving sound source could be produced. Chowning's use of the illusion was a combination of what I see as its two main aspects: motivic and gestural. A motivic use involves the generation of a recognizable 'sound path' that recurs throughout a piece. A gestural use is one in which the illusion of movement through space serves to reinforce or add expression to whatever musical structure does the 'moving' (in much the same way as dynamics might be used). Experience, however, has shown the limitations of this effect. The creation of an illusory sound path which can be heard by a listener within a musical context and recognized as the same the next time it occurs depends on a great many factors, not all of which are under the control of the composer. In the first place, only certain kinds of sound can be used effectively. Continuous sounds, or those with few or slow attacks, cause difficulty. Rapid percussive sounds work better. Use of a Doppler shift can clash with the harmonic or melodic context. In order for the effect to be clearly perceptible almost nothing else can be going on in the music. In the typical set-up using four speakers, the sound system must be perfectly balanced, and the audience confined to a small area in the centre. The listening space must be as acoustically dry as possible, so that the artificial reverberation needed for the effect is not blurred by natural reverberation.

Even though I was compelled by it when I first heard it, I am no longer convinced that this aspect of 'moving sound' holds as great a promise of added expression as other unique aspects of computer music, even when the effect is perfect. Perhaps it has too often been used as a gimmick to dress up an otherwise unimaginative piece. Lately, most people, including myself, seem to prefer to use it perhaps once or twice in a work, and usually in the gestural sense, at some special moment in which the contrast with otherwise stationary sounds adds to its effectiveness.

Innovations in the technology of sound reproduction may someday bring this idea back to the fore. Loudspeakers are in fact the weak link in audio technology today, not only in terms of spatial imaging but in terms of distortion and frequency linearity. A reproduction system which provided true three-dimensional sonic imaging, while not placing any special restrictions on the listener, would mean a tremendous increase in the potential for spatial control as an expressive device.

* * *

The area of computer music that is more difficult to explain and understand is that in which the computer is used as a 'compositional tool'; it is here that the lie is given to the claim that it is 'just' another instrument. Never in the history of music has there been a musical instrument with greater ability to adapt itself to an individual composer's methods and madnesses during composition, learning and facilitating the composer's idea of what is important musically and how it is best controlled. Even though the number of musical decisions for the composer are vastly increased, decision-making rules can be built into the instrument along with his or her personal aesthetic and intelligence, in order to take advantage of the concomitant increase in musical potential.

One of the more powerful innovations in the field of computer science in the last few years has been the introduction of object-oriented programming. This style of programming essentially removes the distinction between 'program' and 'data'. In their place, one has 'objects': user-defined combinations of algorithms, information, and memory. An object is an independent entity which can be interrogated, asked to perform a task or remember something, or even modified by the actions of other objects. Within this paradigm, programming a computer becomes less like traditional programming and more like teaching. This is precisely why it is almost universally used for work in artificial intelligence, where the most problematic of the tasks at hand is somehow to put into the computer the unique and often ill-defined knowledge of human experts.

My experiences with this approach in the field of commercial artificial intelligence led me to experiment with it in music, using the object-oriented programming facility of Bill Schottstaedt's PLA language at CCRMA.[3] I have found it to be an appropriate and powerful method for dealing with the overwhelming amount of data and information which must be handled in making computer music, at all the levels of a composition.

For example, say you have a general synthesis algorithm in which 10 parameters must be set to specific values for dynamic frequency control. In one context where the algorithm produces the sound for a conventional musical phrase, these parameters might be used for control of random and sinusoidal vibrato. Three different types of vibrato might be used for a 20-note phrase, for example subtle, strong, and swelling (i.e. coming in after the note's attack). Changing 10 parameters for each of 20 notes in the high-level program just to get a change among three common variations in vibrato is rather unwieldy, and of course is just the tip of the iceberg when there are likely to be dozens of other parameters as well.

In an object-oriented environment, you might instead have an object named 'vibrato'. You could include in this object algorithms to be executed in response to the symbols 'subtle', 'strong', and 'swelling'. The algorithms would set the levels for all the relevant parameters in each case. You could also include algorithms which might modify the object's behaviour in certain musical contexts, such as 'solo' or 'ensemble'. The high-level program would then have a single list of symbols, clear in meaning, and the knowledge about what you mean by strong, subtle, and swelling vibrato would then be part of the computer's language, there to be used in the future.

The result of this approach is that, as one proceeds, an increasing proportion of the details of the composition are taken over by the part of the composer's own intelligence which is transferred to the computer. The composer is then free to devote more attention to the higher and more abstract levels. This is in sharp contrast to earlier music languages, in which the aesthetic intentions of the composer were often compromised and limited by the narrow assumptions of the author of the language.

I feel it is very important to consider these issues when discussing the aesthetic potential of the computer. The high costs of general-purpose computers with hardware and advanced languages powerful enough to be used for a wide range of computer music has resulted in the design of music synthesizers and control languages based on the premise: "Whatever people are doing most often we must provide the cheapest machinery to enable even more people to do it". This inevitably results in a machine or program which forces the composer to work within the designer's paradigm, spending most of the time trying to figure out ways around the omissions and limitations of the design.

This is ironic considering that the most powerful aspect of computers is their general-purposeness. The key design question ought to be simply: "How much general-purpose audio signal processing and control power can we obtain at a cost which the musician is willing to pay?". I believe this is a critical point which bears emphasizing. In order for the full power of computer music to be realized, we must always strive for maximum generality in new software and hardware. The goal we seek is nothing less than the free expression of our imaginations. No one else should decide for us the best way to get there. All hardware and software intended for musical use should be designed with that in mind. On the other hand, composers must realize that there is no substitute for the study, patience, and care needed to achieve a high level of virtuosity. This is true with any instrument, but is even more true with an instrument capable of such an unprecedented range of expression.

* * *

I am not prone to using composition techniques based on preconceived algorithms. Most of what I have heard or attempted in that direction has suffered from what I consider to be the fatal flaw of containing one (or more) completely arbitrary mapping of algorithmic results to musical parameters. For example, composers often control pitch according to some mathematical model of a physical process or a combinatorial scheme. This procedure usually ignores the resulting harmonies generated, or, in more specific terms, the acoustic interactions of the harmonic structures of the resulting tones. Since the poor listener in this case cannot help but hear the random harmony generated, usually much more strongly than any intended contrapuntal or textural effects, the experience is usually meaningless. Even when the acoustics of the situation are taken into account by the composer, the results of such things as the use of stochastic processes are rarely satisfying.

There is however one area of algorithmic composition which produces surprisingly good results with relatively simple techniques. The results are at times so good, in fact, that they provide some food for thought regarding our concepts of aesthetics and the creative process. This area is that of the application of fractal geometry to music.

'Fractal' is a term coined by the mathematician Benoit Mandelbrot in 1975 to describe a little-explored branch of mathematics concerned with shapes which cannot easily be described by standard geometry and calculus.[4] Such shapes include a great many of those found in nature, including such natural structures as mountain ranges, coastlines, snowflakes, and clouds. These kinds of structure tend to possess a property called 'self-similarity'. That is, viewed at various scales of magnification, the apparent shape remains essentially the same (within limits determined by the physical material). This suggests that these kinds of shapes can be modelled using recursive algorithms. This type of model applies the same basic transformation rules to describe all the levels of detail in a complex structure. For example, one might begin by breaking a straight line into an angle, then each leg of the result into the same angle inverted, then the four resulting legs into the same angle inverted again, and so on. The resulting complex fractal shape would be completely described by the one rule and the number of levels of recursion. Adding random variation to such simple rules can produce very rich natural-looking shapes. The successful modelling of landscapes and the efficient computational nature of fractal models has led them to be widely explored in the field of computer graphics, and applications have been found in many other disciplines.[5]

In the 1970s scientist Richard Voss published the results of analyses showing that the structure of pitch fluctuations in music had fractal properties. If melodic shapes are analysed as if they were waveforms, and 'f' stands for the frequency components of those waveforms, their spectral density always turns out to be proportional to 1/f. Statistically, these shapes are indistinguishable from a kind of noise called '1/f noise', which commonly occurs for example in mountain ranges, transistor noise, traffic patterns on freeways and so on. This type of noise also has fractal properties, such as self-similarity. In music, this property was found across periods from medieval music to rock and roll, and across cultures from the Ba-Benzele pygmies to old Russian folk songs.[6]

Voss also reversed the idea and suggested that music mechanically generated by mapping 1/f noise to pitch or rhythm sounded more pleasing than either doing the same with completely random ('white') noise or $1/f^2$

('brownian') noise. Indeed, melodies generated this way do sound more reminiscent of composed music than those generated with other kinds of random functions. I am not sure about 'pleasing', but it definitely has the right kind of basic shape and sounds somehow 'intentional'. What the mind hears is the correlation between notes, which is of course a property of most music. With white noise, each new value in a series of random values is completely independent of the previous and the next. With 1/f noise, each value is correlated with the values on either side, with the influence decreasing as one gets further in the series from the value in question.

At the time that I first read Voss's work, I was looking for a way to generate a natural-sounding random vibrato for synthesized voices. Using a random number generator (white noise), the vibrato did not sound at all natural. However, a 1/f noise vibrato sounded perfect. Upon reflection, this result seemed logical, since the elastic fluctuations in the vocal chord muscles would naturally produce a correlated series of positions. They would obviously not jump suddenly from one random degree of contraction to another. I went on to experiment with some simple uses of 1/f noise in melodic generation.

Since then, interest in fractals has blossomed in a great many fields. I decided to try to use true geometric fractals as a prime compositional aspect in one movement of the five-movement ballet *Invisible Cities* (1985). Details of the mathematics are beyond the scope of this book, but the interested reader may refer to the works cited in the notes to this chapter, or to a more technical paper which I have in preparation. (Suffice to say that I used the Weierstrass-Mandelbrot random fractal function.) Melodies were generated using the fractal values to select notes from composed harmonic modes. Certain notes in the modes were given weightings which increased the likelihood of their selection. In addition, variables were specified which occasionally forced the melodic pattern to hit a particular note at a particular time, to satisfy the basic structure of the piece. The dynamics of the melodies were also determined by the same type of function, but one based on a different random series. (This type of fractal is completely specified by a handful of random values and a few other parameters, such as fractal dimension.) In some cases, repeated notes were avoided by a simple rule-based scheme. The rhythmic structure was also related to the melodic fractal function.

The result of this relatively simple scheme, within the larger context of the piece, was frequently astonishing. There are some rather hair-raising extended sections which sound like a good human jazz or rock improviser, in melody, rhythm, and dynamics. It is quite uncanny, and would imply that here might be a clue to how our minds create and relate to music. Perhaps the structure of our brains is the result of a fractal process, and this is reflected in the way we process and store information, in the ways that we choose to express ourselves, and in the kinds of external structures that we prefer to deal with. The fact that fractal properties are common to music across such a wide range of cultures and times at least suggests that there is physiology-based common ground for musical aesthetics.

8

Computer Music Language Design and the Composing Process

Barry Truax

The use of the word 'language' in the title of this book, referring to the structure of musical language, is fundamentally different from its use here with reference to the structure of a computer music system. Conventionally, computer languages are programming languages, whether on the level of machine operation instructions or the so-called 'higher-level' languages through which more abstract relations can be expressed. I shall refer to *computer music languages* more broadly as the complete software system that allows a user to interact with data and/or sound. Since sound itself may be represented in a computer music system as data (either literally as digital numbers describing the sound, or more abstractly as control parameters), we can generalize the definition of a computer music language to mean the software system that allows the user to deal with compositional data at any level.

What I would like to examine is the relation between the computer music language and the musical language which is the product of its use. The key to this examination, in my opinion, is to focus not on the diverse styles of computer music but on the composing *process* and how it is affected by the design of various types of computer music software. I will argue that the structure and behaviour of music software provides a framework – a set of concepts and tools for their use – within which the music is conceived and realized and, moreover, that the computer affords new possibilities for organizing and manipulating sound (whilst allowing for the imitation of traditional methods). These new frameworks allow the composer to think differently about sound and to evaluate the results in terms of their musicality. In a recent book I have argued that the importance of the computer for music lies both in its digital representation of sound (and the manipulation techniques which that form allows) and also, equally importantly, in its potential for designing the compositional process.[1] In fact, I believe that it is these new processes afforded by computer music software which are the most influential in changing the language of contemporary electroacoustic music.

The structure and behaviour of computer music systems

How do computer music systems allow the composer to think differently about sound? Part of the answer comes from the structure of the system, and the rest from its behaviour when in use. The structure of the system will be reflected in the answers to such questions as:

1. What can be specified and controlled?
2. What concepts define the different components of the composition, that is, how are the data grouped into different organizational levels?
3. What procedures and methods of organization are available, and at what levels do they operate?

A fundamental trait of the practice of electroacoustic music is that the composer composes the sound itself as well as the structure in which it appears. A computer music system gives symbolic and numerical representation to both sound and structure. The above questions attempt to reveal the data structure and procedures for controlling it found in any given system. Just as words and grammatical structures allow different ideas to be expressed in natural language, so too the structure of a computer music system constrains and facilitates the composer's thoughts in particular ways. For instance, each sound synthesis method allows various micro-level and macro-level parameters of the synthesis to be specified, sometimes as constants, in other cases as time-dependent functions. The precision of such control is one of the main advantages of digital synthesis over analog, voltage-controlled electronic synthesis. But anything given over to the composer's specification may influence larger-scale compositional organization on the basis of conceptual and perceptual considerations and may result in a new kind of music.

The grouping of parameters and data is just as important as what each represents. In musical perception it is often the way in which various parameters combine to form a whole that is more important than the pattern of organization of any one parameter. In a computer music system, the grouping of data into larger units such as a sound-object, event, gesture, distribution, texture, or layer may have a profound effect on the composer's process of organization. The challenge for the software designer is how to provide powerful controls for such interrelated sets of data, how to make intelligent correlations between parameters, and how to make such data groupings flexible according to context.

Parameter groupings are just one instance of the introduction of hierarchic levels in a system. Musical structure can usually be thought of as multi-levelled, and therefore a computer music system that can address different levels of musical structure effectively will be a powerful compositional tool. And powerful tools lead the creative imagination in new directions. However, the introduction of each level of hierarchy implements a specific musical–acoustic model and 'prejudices' the system with a specific hypothesis about how music is structured. The higher the level the greater the prejudice. The lowest level, for instance that of sound samples, is not restricted to music at all (since it can describe any sound), but as each higher-level concept is introduced it contributes to a specific musical model and requires more music-specific knowledge to be implemented in the system. Small wonder that powerful compositional systems are few in number (given the difficulty of formulating the knowledge on which they are based) and often regarded as idiosyncratic (given the specific musical worldview they must necessarily implement). Conversely, general-purpose systems, showing less prejudice because they incorporate less knowledge, are notoriously harder and slower to use since the knowledge they lack (about musical structure) must be embodied in the data provided by the user.

The above discussion has attempted to show in a general way some of the implications of the structure of computer music languages for the way composers can think about and structure sound in a musical composition. The most general answer to the question "Why use a computer?" is that it is suited to organizing complexity. If a composer wishes to deal with a complex problem, whether it is an interactive live performance, or a particularly subtle control over sound synthesis, or the multi-levelled control of music parameters, then a computer music program is a powerful way to deal with that problem.

If the structure of a computer music language determines how music can be conceptualized and represented (in other words, what is potentially 'thinkable' by the composer), then the behaviour of the system determines how the composer interacts with the possible data structures (in other words, what is actually 'thought' by the composer). In 'behaviour of the system' we may include all questions of how that system functions when interacting with a user, such as:

1. What is the form and modality of user input and program output?
2. How are the data structures at various levels represented to the user, and how efficiently can changes be made to them?
3. Does the user have access to sounding results at any stage and with what degree of speed and accuracy?

The answers to these questions reflect what I call the 'communicational environment' within which the user works,[2] and it is my contention that the nature of this environment influences the musical output to a significant extent. Its effect on the 'learning curve' that characterizes the system is obvious. Acquiring skill with any tool is well known to be dependent on the kind of feedback given to the user about the actions that have been taken. So too with a computer music system. The more readily and effectively the results of the user's input are seen and heard, the more efficient is the development of control skills. Therefore, real-time interactive systems are clearly superior in optimizing the learning process.

A more subtle aspect of system behaviour relates to the strength of organizational methods that are available. I have elsewhere outlined the dialectic relation between the concepts of generality and strength as applied to computer music systems.[3] A system that uses weak methods and relies on a great deal of user input can produce a wide range of output because of the generality of the methods used. On the other hand, the use of strong methods which require less user data results in more restricted, but well-formed output (or at least as well-formed as the extent to which music knowledge is implemented in the methods used).

My hypothesis is that the use of strong organizational methods in a computer music system will make the potential range of musical output more readily accessible to the user and that, if accompanied by sounding or score output, those results can be effectively screened for their musical value. The behaviour of such a system will encourage musical exploration, instead of a reliance on what is already known to work. Although the same output could be generated, at least theoretically, with a general-purpose system, it would be more cumbersome, and the user is less likely to discover the path leading to the final result. Powerful tools make complex operations accessible; the creative mind is less burdened with details that are not deemed important and free to explore many pathways before deciding on the most promising one.

In the interactive system, the program participates in the compositional process which therefore can be described as computer-aided composition; on the other hand, in the general-purpose system, most compositional decisions are made before the program is used, and hence the process can be described as computer-realized composition. In the latter approach the problem is, given the musical idea, finding the means to realize it; in interactive systems one is given powerful tools, and the problem is to find the music. More often than not, that music will be discovered not imposed, revealed not asserted.

The language of computer music is not independent of the language that is used to create it if the computer participates in the compositional process. The inherent quality of software is its flexibility, and therefore there is no end to the extensibility of computer music languages to incorporate new compositional concepts. In practice the real limitations are those that are human-imposed, not the technological ones. Arbitrary restrictions (for instance, to fixed timbre per musical voice, to rhythm expressed in metrical units, or pitch in equal-tempered scales) abound in computer music systems (and the popular digital synthesizers which are their offspring) and clearly these restrictions constrain the musical output. It is certainly not the first instance of a new technology being used to reproduce older forms and concepts, but it is certainly one of the more ironic ones, given the inherent flexibility of the medium. Ultimately, a computer music composition reflects the musical knowledge embodied in both the software system that produced it and the mind of the composer who guided its realization. The interaction between these two bodies of knowledge is the essence of the creative musical process.

Types of software approaches

Some insight into how computer music language design affects the composing process can be obtained by examining a system designers' documentation of the language; a different point of view is seen when a composer documents how a particular composition was realized with the system. Most of the available literature falls into one or both of these categories, and in each case it is usually the structure of the system or composition that emerges most clearly from the discussion. How the behaviour of the system influenced both the composing process and the musicality of the result is seldom made explicit, partly perhaps because these questions appear to be too subjective, and partly because the active, cognitive nature of process and musicality makes them difficult to document and explain.

The reliance on structure, whether of software or music, as a basis for the design of a computer music language creates other problems. For instance, it is not clear to what extent computer programming concepts, techniques or languages correspond to their equivalents in music. Given the poorly defined nature of musical tasks and the open-endedness of creative thought, it is difficult to ascertain what programming structures are most appropriate for music. Computer music languages are often designed for their power and efficiency (if not novelty) in computing terms. The composer is left to realize a musical idea with confusing or inappropriate tools using the given programming language – a formidable task which requires a great deal of trial-and-error testing.

On the other hand, if we start with what we think we know about music and infer what program structure best facilitates its production, another kind of problem arises. It has been conventional, largely through work in music theory and musicology, to rely on the score as the musical artifact which is the visible result of the composing process. In Western music the score represents a structure separated from the sound through which it will be realized. Therefore, it is hardly surprising that the principal model of computer music software is the *score editor*. The problem is that the score is a description of the surface level of the resultant composition, not a guide to the musical thinking that produced it. By modelling music as a score to describe structure and instruments to perform it in sound, we also model composition as the assemblage of the notes of the score and the detailed specification of the sounding instruments.

Another way to understand the problem of the score as the focus of computer music composition is to ask, given the resulting score (assumed to be musical), how it could have been produced. There are numerous processes, at least theoretically, which could have resulted in the same output. The answer provided by the general-purpose system is to make assemblage of the parts of the score the object of the process, not the description of its result.

The distinction is the same as that between a 'creative writing' program and a word processor. The latter may facilitate the assemblage of the words and sentences, but it provides no particular tools to organize the thoughts they express. However, with the computer as means, with its ability to execute procedures, it is possible to think of the machine as more than something which assembles the result of creative thinking.

The score editor, as a general-purpose tool in computer music, still has considerable usefulness. With the traditional MUSIC V model,[4] each note of the score can have a long list of parameter values associated with it which apply to the defined instruments. Hence the language permits very detailed control over the synthesis parameters, proportionate to the amount of data supplied by the user. One of the best documented score editing systems functioning in an interactive composition environment is that of Buxton *et al.*[5] Sophisticated graphic means are used to enter, edit, and orchestrate musical notes in a variety of representational modes. One advantage of such an approach, even though it embodies a fairly traditional score-editing-as-composition model, is that composers who are used to conventional methods have almost immediate access to the power of the system.

Buxton has further shown how powerful tools pertaining to score editing can correspond more closely to compositional thinking than simple note-by-note procedures.[6] First of all, defining what he calls 'scope', or what in my PODX system[7] is called 'conditional editing', provides a powerful musical tool. It refers to the specification of a subset of the score based on common parameter values (e.g. all pitches in a certain range with a given timbre); such subsets can be manipulated as a unit for a specified transformation. Likewise, powerful operators that can change a given set of data systematically (whether in a predictable or controlled random fashion) may also enact a transformation suggested by a musical evaluation of the previous version. The important ingredient is that the program offers data manipulation tools which correspond to the perceptual and musical concepts of the user.

Besides operators that act on already existing data, flexible methods for generating the data are also useful. Buxton uses the concept of 'instantiation' to refer to a basic pattern or grouping of data which can appear in a number of variant forms.[6] Within a hierarchic representation, changes to the basic definition of the pattern will automatically cause changes to each of its 'instances' in the score. A simpler case of score automation is to provide a variety of selection methods by which data can be chosen, ranging from deterministic algorithms (e.g. linear or exponential change, sequences and their permutations) to stochastic (e.g. aleatoric, weighted probability distributions, tendency masks, 1/f selection). Such methods relieve the user of redundant input and simultaneously provide for simple patterns which may create the basis of a musical gesture.

Contrast these kinds of explicitly musical tools with the total reliance on note-by-note and parameter-by-parameter specification found in much of the commercial music software for microprocessor-based systems. The difference is between software design actively assisting the compositional process in new ways and its simply imitating the most superficial aspect of the traditional method. It is not, therefore, the representation of musical structure as a 'score' – whatever that may mean as a set of data – that is problematic in computer music systems. With more powerful compositional methods, the score (or 'file' as it is more commonly and neutrally called) is the *result* of the organizational process, not the focus of activity during the process. Files are manipulated and calculated under program control according to higher-level decisions made by the composer; the score is merely a convenient representation of the data being used.

In the PODX system (and probably in most other specialized compositional approaches as well), it is only the so-called 'basic' level of the system that incorporates the general-purpose score editor where each event or timbral parameter may be independently altered. The 'specialized' programs within the system, each of which implements a specific compositional model, characteristically treat score parameters as dependent variables. That is, more than one parameter at a time is determined by a higher-level process. It has been one of the fundamental musical problems of this century to balance the analytical approach, where each musical parameter can be organized independently of every other parameter, with the basic fact of musical perception, that all features of a perceived event or group of events contribute to a single gestalt. Musical perception seems more oriented toward discerning pattern at different hierarchic levels (i.e. patterns of patterns), rather than parallel processing of individual parameter patterns.

Types of approach other than score editing include use of musical grammars[8], stochastic processes[9], and automated or rule-based composition, such as that of Ames[10], Englert[11] or Koenig[12]. More recent examples of interactive, high-level composition languages are the FORMES system[13] operating at IRCAM and the HMSL system[14] at Mills College. Composition involves activating certain rules or processes, usually to control syntactic structures. Since most of the control principles do not derive from conventional musical techniques, the composer is frequently exploring the musical potential of the use of such rules or processes. All of these approaches create new types of compositional methodologies, if not actual musical organization and meaning.

Chadabe has made the very useful distinction between systems employing 'memory automation', of which all of the data-intensive, score-oriented systems described above are examples, and those more concerned with 'process automation'.[15] Compositional thinking in the latter case is much more oriented toward systems because one is designing interrelated processes, not data. Such systems are frequently designed for live performance, particularly if they involve performer input which controls the behaviour of the system. Chadabe calls his own approach 'interactive composing' because each performance depends on how the composer-performer interacts with the rules of the system.[16] Martin Bartlett's work frequently involves an analysis by the software of a live performer's output and a response in a rule-governed manner.[17] Doug Collinge's MOXIE language may also be mentioned as a means of specifying actions and organizing them in time.[18] In each case the composer is setting up a rule-governed system with a variety of possible outputs, instances of which will emerge through the interactions that take place during live performance. The challenge is to ensure musicality *indirectly* through an effective set of rules and constraints that can operate within a dynamic and only partly foreseen performance environment.

Automated, interactive, and process-oriented performance systems are all examples of how procedural knowledge (as well as stored data) can be integrated within a computer music system. Each extends or even redefines the compositional process, and each has the potential to create new musical languages. Each is exploratory in that it allows new structures to emerge from a fixed set of rules, structures whose musicality is a test of those rules. Such systems characteristically are in a constant state of expansion as their use (through the mistakes as well as the successes) suggests further development and even more effective means of generating and controlling complexity.

Specialized compositional approaches in the PODX system

As an alternative to the general-purpose score editor and as a significant extension to its limited musical power, the higher-level, or 'specialized', programs in the PODX system[7] all have in common the practice of working with a particular compositional *concept* which the program is able to translate into an output file. The user can specify and change the control structure interactively, hearing a sounding result through the polyphonic synthesis system. All of the PODX approaches are examples of 'top-down' compositional processes, although the user also has access to lower-level details, either in testing individual sound-objects, or through conventional score editing.

Two characteristics of the higher-level compositional program are that the user is involved in specifying control structures corresponding to *potential musical entities* (and therefore may be oblivious to the score representation which is the result of those specifications), and secondly, that the control structure generates relations between *more than one score parameter* at a time. As a result, a relatively small amount of input data is required compared to the output that is generated, and any *one* change in that input results in *many* changes in the output. Because the output is often so complex that it cannot always be predicted, feedback of sounding results (even if imperfect in sound quality) is important for the user to evaluate the musicality of the output. Such evaluations prompt further optimization of an intermediate structure, or even complete redefinition of the goals.

Some of the higher-level concepts found in these programs are density, sound distribution, timbre mapping, structure variants, the use of masks to define vertical sound density, spatial trajectory, and trajectory mapping. Although most of these concepts are foreign to traditional compositional thinking, they have a close relationship to electroacoustic compositional practice, and in most cases extend the level of structural control that is possible within the analog studio. These concepts are also designed to correspond to how sounds are perceived, patterned, and represented in the mind. In fact, one of the most important distinctions in high-level approaches is between those that allow completely abstract relations to be formulated, and those that correlate strongly to perceptually-based relations such as forms, shapes and patterns. Let us examine some of the PODX approaches in more detail.

The earliest compositional model, found in the POD6X program and its predecessors (POD4, POD5 and POD6), involves the user in creating distributions of sound.[19] In these programs the user specifies a frequency–time mask which delimits the frequencies by allowing only those inside the mask to be chosen; the user also controls the average density of sound-events throughout the mask. From a Poisson function a stochastic range of durations is generated which cluster around that average. The other major task of the user is to specify and control the mapping of timbral entities (i.e. sound-objects) onto the events of the distribution. Various selection methods are provided, including tendency masks for time-dependent selection of timbre. Other aspects of POD6X include performance variables for altering and optimizing the interpretation of the score, as well as the possibility of generating variants (including random ones) of a given structure. Although stochastic rhythms may be generated through use of the Poisson distribution, other distributions, including quite deterministic and even metrical rhythms, are also possible.

The maximum amplitude and object number associated with any one event is determined by separate selection processes which may be correlated by the user with the frequency–time structure. The score is sufficiently ambiguous to allow various interpretations via use of performance variables, particularly in the area of rhythm and the specification of frequency modulation parameters.

Another quite simple approach is found in the PDMSKX program which was originally designed during the composition of my piece *Arras*.[20] Here a tendency mask is used to form the boundary of simultaneous events. The result, once the frequency spacing of the events is specified, is a precise control over vertical density. Contoured frequency-clusters and similar 'sound blocks' are easily specified. In the resulting score, the simultaneous specification of entry delay, duration, and frequency having been achieved, timbre may be controlled via score editing.

The most recent approach, found in the PLOTX program, is based on spatial trajectory and its correlation with timbre.[21] The user plots a path on a screen for the sound to follow within a two-dimensional acoustic space. The sound is assumed to be more or less continuous, travelling either with constant time between points or at constant speed. The trajectory is realized with a set of overlapping component sounds, each of which has its own timbral specification. The calculation of the score determines the time structure as well as the parameters necessary for distance perception and stereo spatial position. Only timbre is left to further specification.

In the simplest case, timbres are specified with the aid of the score editor. However, two more sophisticated methods are also available within PLOTX. One combines the trajectory with an existing user file and its timbral structure. This can be thought of as *timbral mapping* because the timbres in the user file are copied to the events in the trajectory file. In the second method, the trajectory is applied to each and every event in the user file allowing the possibility of *overlapping* trajectories.

These methods achieve the most complex possibilities of mapping timbre onto rhythmic and spatial patterns, and they are the closest to having all score parameters determined by the same process. The compositional approach is also the most specifically 'environmental' in nature and hence possibly the closest to traditional electroacoustic music practice, but with much greater control. Like the other higher-level methods outlined here, they change the composing process by allowing the composer a level of control that closely corresponds to the perception of musical gestures and patterns. The programs involved are equipped with procedures to generate the required low-level data that otherwise would be too complex for the user to provide. Synthesis and score data are still open for the user to change when needed, but their direct specification is not required. The conceptual power of the procedures encourages exploration, as its cost to the user in time and energy is slight in comparison to the richness of output. If new forms of musical expression are to emerge from computer music systems, they may well be discovered through use of such tools.

Yet, despite the increasing sophistication of existing compositional software, it still lags behind the advances in sound synthesis and processing that are rapidly appearing. Innovative software for controlling what can be produced is slow to emerge, perhaps because developmental systems seldom allow experienced composers to work with the new processes and to suggest how control languages should be developed. The rush to produce the marketable package frequently commits a new system to fairly rigid methods of control, and proprietary interests restrict user modification and adaptation. Powerful high-level software with a strong correlation to musical perception is difficult to find. Fear of such software being 'idiosyncratic' to one composer (as if specialized musical approaches cannot be shared, or general-purpose systems preclude use of specialized ones), combined with a lack of fundamental research into musical cognitive behaviour, are reasons that readily come to mind to explain the lack. Still, we have moved some way from Risset's observation that in the early days of *musique concrète*, the sound could "only be transformed in ways that are rudimentary by comparison with the richness of the material"[22]. Refined control over sound material has increased dramatically in recent years to the point where few composers with experience in the analog studio have resisted some form of exploration with digital techniques. However, one might justifiably rephrase Risset's observation by saying that today languages for the musical structuring of digital sound-material are less sophisticated than the methods for generating that material. We are only now beginning to discover the expressive musical potential of the digital sound-world we have created.

A compositional case history

To conclude this chapter, I would like to illustrate the points I have been presenting with an account of the composition of a recent work of mine, *Solar Ellipse* (1984–5) for four-channel tape. I doubt that reading any account of the compositional potential of a computer music system can convey an understanding of how the composing process is shaped by it as effectively as working with such a system. If a composer works out a musical structure and then engages a computer system to realize it, factual information about the realization techniques will satisfy the curious. But if, on the other hand, the computer is equipped with programs that actually participate in the compositional process, and the end result is evolved rather than predetermined, then the experience is much harder to explain. My present account will suffer from the same difficulty, but possibly it may provide some insight into the relation between process and resulting structure.

Solar Ellipse was designed to complete a four-work cycle of pieces, entitled *Sequence of Earlier Heaven*, that includes a pair of computer synthesized pieces and a pair of pieces for live performers accompanied by tapes derived from sounds recorded from their instruments (*East Wind* (1981) for recorder and tape, and *Nightwatch* (1982) for marimba and tape). One of the computer pieces is *Wave Edge* (1983) which was the first to use the trajectory concept extensively and *Solar Ellipse* was to be its 'mate' using complementary and contrasting techniques. The central trajectory and environmental image of *Wave Edge* is a right-to-left movement of waves breaking along a shore where the sound of the water remains fixed in direction while the crest or edge of the wave moves laterally. In *Solar Ellipse*, the complementary image is that of fire, and the trajectory that of revolving motion, specifically the epicycle where a spinning sound-image revolves around an elliptical orbit. The compositional goal with this piece was strongly determined by its having to fit into the larger cycle of works.

I began by exploring the smallest unit of the piece, a sound that would appear to spin. My previous work with trajectories had been on a larger scale, but using the same method of specification I was able to design a small circular pattern lasting a quarter of a second. Computer synthesis frequently deals with long sustained sounds and slowly evolving textures, and for that reason I wanted to explore alternative sound structures that would give the impression of speed. The smallest trajectory that sounded convincing comprised six events having entry delays of 5, 5, 4, 2, 4, 5 centiseconds. The reason for the unequal time values was to give the impression of constant speed; the sound appears to move slightly faster as it comes closer and slower as it recedes. In addition, a small Doppler effect was added to raise the frequency as it approached and lower it as it receded. Despite the brevity of the pattern, such fine detail adds a sense of realism to the sound.

The basic spin unit was then put into a larger trajectory pattern where it would travel around an elliptical 'orbit' defined by 18 spatial points. The resultant file lasts 4.5 seconds. Each of the 18 positions in this large trajectory is assigned a related timbre. For instance, those in the right quadrant use a complex modulating wave and those in the left a complex carrier wave.[23] Tighter versions of the larger trajectory incorporating fewer spatial positions and lasting 3.5, 2.5, and 1.5 seconds respectively were also constructed. However, a file with eight repetitions of the large epicycle pattern, lasting 36 seconds, became the source of most other patterns used in the piece.

One variation of the epicycle pattern was achieved by prefixing and appending a spiral trajectory that begins with a broad sweep and tightens into the dimensions of the epicycle. As it tightens, its frequency rises in a glissando to that of the epicycle, and after several 'spins' of the latter the spiral reverses itself, as does the glissando. Four versions of this pattern were created, centred in turn on the 4.5-, 3.5-, 2.5- and 1.5-second epicycles. Since their overall durations are different, the shorter ones were given a delayed entry when all four were mixed together, thus preserving symmetry. This pattern was used in the piece to mark the midpoint of the work (5′24″).

Although the epicycle pattern is interesting, it cannot be sufficiently varied by itself to maintain interest over a longer period of time. Therefore the epicycle was used as a building block in constructing four larger forms as follows:

1. The 4.5-second epicycle was lengthened eight times with two versions an octave apart, providing a 'slow motion' ellipse on the pitch G.
2. A pyramid shape was constructed of five layers, being 9, 7, 5, 3, and 1 repetitions of the 4.5-second epicycle in ascending octaves. In other

words, after each repetition the next higher octave is added until five are present; the process is then reversed.

3. A parallelogram was constructed with the same five octave layers, each comprising four repetitions of the epicycle and beginning with the highest octave. In other words, after each repetition the next lower octave is added, but when the fifth one enters, the highest drops out, then the next lower, and so on until only the fundamental is left.
4. A constant-frequency pattern was constructed from the four original epicycles. This consisted of epicycles with durations 4.5, 3.5, 2.5, 1.5, 1.5 seconds and the retrograde of that order. The effect is to create a speed-up in the epicycle pattern followed by its slowing down again.

Each of these four patterns was generated as loops which run the total duration of the piece. A technique for looping the material stored on disc was invented. The scores for each layer are edited to include a fade-in and fade-out at the beginning and end. The sound from the beginning is then mixed and overlapped with its end portion; only a fraction of a second's overlap is needed. The digital 'splice' to form the loop is inaudible when this technique is used.

The resulting durations of the four patterns are in the proportions of 8 : 9 : 8 : 6. If synchronized, they will repeat as a global pattern after 72 time-units, or 324 seconds. It was decided that the duration of the entire piece would be 144 repetitions so that the phase pattern of the various elements would be heard twice; therefore, the work lasts 10′48″ with symmetry about the middle. In fact, cyclic behaviour, which incorporates mirror inversion, is the generating form for the work, from the smallest spinning quarter-second unit to the largest cycle which is the duration of the work itself. Within the large cycle, each component cycle repeats at three different rates, thus providing considerable variety.

The *Sequence of Earlier Heaven* cycle of pieces is also based on complementary or 'yin–yang' pairing. The cycle consists of two pairs (instrument-based and computer synthesized) of two pairs of works each, one pair contrasting the recorder and marimba (symbolizing wind and earth respectively, and literally contrasting air-through-wood and wood-over-air in their respective construction), the other pair contrasting linear wave motion with cyclic wave motion (symbolized by water and fire respectively). To extend the principle further, both *Wave Edge* and *Solar Ellipse* incorporate contrasting timbral families based on vocal formants and noise spectra. In terms of FM, this simply means using multiple harmonic carrier waveforms to produce formants around each harmonic, and complex modulators incorporating a controlled amount of randomness added to a sine wave for the noise spectra. Pitched material reinforced by resonances contrasts with pitched material that has broadband components.

Therefore, each of the four global sound patterns, as documented above, was duplicated with noise-band timbral structures, resulting in eight stereo files each lasting the duration of the work. The four stereo pairs were recorded on the channels of an eight-track tape in the order listed above. Digital reverberation was added during the recording, and for one of the noise-band files, comb filtering modified the timbre as it repeated throughout its duration. The number of samples of delay in the comb filter was systematically altered to create a cyclic pattern of extremely short delays (with resultant high-frequency cancellation) proceeding to long delays (with low-frequency cancellation) and back during each half of the piece.

As might be imagined, each of the two eight-track tapes (one being the formant version and the other the noise-band version) were extremely rich when heard with all tracks combined in stereo. The inner cycles of the loops going in and out of phase provided dynamic interest, but the overall energy level remained at a maximum. In order to provide a more varied dynamic shape to the entire piece and expose the interaction between layers more effectively, I decided on a stereo mixdown scheme that would include all possible combinations of the four stereo tracks on each tape, as follows:

Tracks			
1 & 2	1 0 1 0 1 0 1 0 1 0 1 0 1 0 1		1 1 1 1 1 1 1 1 0 0 0 0 0 0 0
3 & 4	0 1 1 0 0 1 1 0 0 1 1 0 0 1 1	S	1 1 1 1 0 0 0 0 1 1 1 1 0 0 0
5 & 6	0 0 0 1 1 1 1 0 0 0 0 1 1 1 1	S	1 1 0 0 1 1 0 0 1 1 0 0 1 1 0
7 & 8	0 0 0 0 0 0 0 1 1 1 1 1 1 1 1		1 0 1 0 1 0 1 0 1 0 1 0 1 0 1

In the above scheme, a 1 means that the track is present in the mix, whereas a 0 means that it is not. The duration of each unit is about 20 seconds. Note that only the lowest-pitched material (on tracks 1 & 2) is heard at the beginning and the high-pitched material (on tracks 7 & 8) at the end. The middle tracks use the pyramid and descending octave patterns. All tracks are heard at the midpoint separated by the spiral (S) described earlier. The above scheme applies to the formant version only; for the noise-band version, a mirror inversion of the scheme was used. Its mixdown therefore starts with the high-frequency material alone and ends with the low-frequency material (the sound of which has a breathy formant quality often compared by listeners to a rough choral sound or steam whistle). When the two mixdowns are combined, their mixing patterns are therefore exactly complementary. Long digital delays with feedback give the entire mix an exaggerated spatial feeling commensurate with the large-scale gestures implied by the timbral textures.

Although the inexorable unfolding of the sound patterns I have described could have sustained the entire work, I decided to add two other somewhat 'foreign' elements to it. To push the analogy of planetary orbit further, one could imagine the elliptical trajectory crossing the path of another object. These two elements are a very high-frequency sound and a faster, more complex timbral ellipse pattern. This appears first at its normal speed, then in a faster version that culminates in a sweeping downward glissando as it slows down (recalling the glissandi in *Wave Edge*). During the second half of the work, the exact same structure is encountered in reverse (through tape reversal).

Solar Ellipse can be heard as an extension of *Arras* on several levels, the main one being the constant spatial flux of the components, another being the larger timbral palette. However, both are compositions whose macro- and micro-forms are entirely based on timbral considerations. In the newer work, though, there is a close connection between the micro-level spinning pattern and the macro-level cyclic behaviour (though it should be noted that the work does not return to its beginning but rather to its 'timbral mirror' where formant and noise-band versions are reversed). Both works are based on a constant fundamental (approximately the same in fact), but in *Solar Ellipse* the presence of the six-octave reinforcement of the fundamental gives it a much stronger tonal centre. The gravitational 'weight' of that centre is contrasted by the ceaseless movement and permutation of the elements that occur above it – a flux that counterbalances the stasis of the fundamental and produces through their interaction the 'hypnotic effect' reported by many listeners.

One may ask how the structure and behaviour of the PODX system influenced the composing process that took place with this piece. The work was composed in the same order as that in which it has been described above; that is, starting with the smallest 'modular unit' and incorporating it into the next higher level. The PODX system participated in each of these steps, except for the final mixdown stage which achieved the final structure. In other words, the piece was constructed from the lowest level up in a hierarchic manner, but within each level the composition was top-down. The system of PODX programs allows the user to generate patterns whose lower-level details are computed (such as the trajectory pattern), and then to combine these patterns into larger forms. The interactive nature of the system allows each of these levels to be tested and optimized independently before proceeding to the next level.

Had the final structure of the work been conceived from the outset (as was that of *Arras*) and its component levels worked out in a top-down manner – the reverse order to how they were in fact realized – the resulting piece would have been very different. The compositional solution found at each level of work on *Solar Ellipse* was the result of a problem posed by the next lower level. For instance, the mixdown scheme was invented because of the uniform richness of the eight component tracks; the eight tracks themselves were created to allow a phase pattern to emerge from the shapes of the four kinds of loops; the four larger shapes of multiple trajectories were invented because of the repetitiveness of the individual epicycle; and the epicycle itself was invented to give dynamic shape to the quarter-second spin. Therefore, it was the power of the system both to provide precise control of detail and to facilitate the incorporation of each unit within a larger pattern that allowed the work to emerge in the form that it did. In retrospect, further program aids to facilitate work in this manner could be envisaged, and, in fact, it is this kind of experience, derived from actual compositional needs and practice, which is the driving force behind PODX system expansion.

Conclusion

The language of computer music today is at a highly evolutionary stage, having, in some cases, elements which reflect earlier electroacoustic or instrumental music practice while creating, in other instances, a new world of musical expression. I have argued here that behind the new 'styles' being created lie new compositional processes and new ways of thinking about sound which can be traced to the specific computer music software that is being used. Of course, it is not just the software that is producing the change, but rather the interaction between it as a source of procedural musical knowledge and the composer's musical sensibilities. Of interest are the systems which the composer can treat not just as a new realizational tool for previously worked out ideas, but as an intelligent partner in the compositional process. And if some of those partners still seem less intelligent than we would like, it is only because we have not invested them with sufficient musical knowledge or powerful enough musical procedures.

The key to progress in software system design seems to me to lie in the creation of higher-level compositional languages that allow the composer to work with sound material and structures at a variety of levels, with significant interconnections between levels. Furthermore, the organizing concepts embodied in such software must correspond to perceptual and cognitive musical structures, supported by algorithms capable of realizing all lower-level details. In such systems, it should be possible to evolve as well as predetermine compositional goals, with aural feedback at every stage. And finally, knowledge gained about successful compositional strategies must be fed back into the system in the form of expanded resources.

The fact that we do not have such an ideal system today, nor do we see enough work being directed toward such goals, lies in poignant contrast to the fact that many of us have glimpsed its potential.

9
The Mirror of Ambiguity

Jonathan Harvey

When composing, the normal course of action is to imagine a sound, sometimes check or adjust it at a piano, then write it down: notate it. Is there any essential difference between such traditional ways of working and the composing of computer synthesized music? At the terminal a sound is also imagined, tried out, adjusted and then saved in program language notation: a very similar procedure.

Once the long preparatory hours of grey programming tedium are over (something like ruling your own manuscript paper) there is an indescribable excitement about working with a computer system. It is quite a different feeling from that of traditional composing. It is, I think, the sensation of confronting a *mirror*. One is searching for the final form of the sound, as it will be played at concerts, whereas in traditional composing one is searching for the final *notation* that will *lead* to the desired, still imaginary sound. Adjustments are made in response to a sound fed back through the system's speakers. We ask ourselves "What number is needed to turn that sound, that decay, that fifteenth partial, that amplitude vibrato accelerando into something recognizably right?". We transform our response into numbers. We quantify precisely. But the response is essentially emotional. Words, indeed philosophy as a whole, cannot approach with objective precision such vitally significant areas of human behaviour. We don't like the sound: we may be able to rationalize a little about our dislike – "It's too like a washing machine or a belch" – or if we have a clear imaginary vision (not always the case by any means) we can say it's not what we were looking for because of some mistake; more usually we know it's not what we were looking for, yet cannot say what we *were* looking for.

At the terminal we must respond to that emotion with numbers. Every detail of 'how much' must be supplied to the dumb computer. This is very different from working with musicians, to whom we say such things as "more lyrical", "more aggressive", "*più nobilmente*", and they immediately adjust a hundred parameters of spectrum, speed, pitch change, and so on without thinking. The musicians react at a much higher level, where it is not necessary to descend into thinking, into deciding whether an amount should be 2 or 3; they just 'feel it'.

But it is in this very inconvenience of having to look so intimately inside an emotion, and calculate so accurately, that the extraordinary fascination of working with computers lies. We seem to be contemplating an impudently truthful mirror which asks appositely precise questions. We change a number; we react emotionally to its effect. We change another number; it has a subtly, intriguingly different, emotional effect. We change another number . . . and so it goes on many hundreds of times, this back-and-forth mirroring of subjective and objective. A kind of mutual adaptation is reached.

> What human music is, anyone may understand by examining his own nature. For what is that which unites the incorporeal activity of the reason with the body, unless it be a certain mutual adaptation and as it were a tempering of low and high sounds into a single consonance? What else joins together the parts of the soul itself, which in the opinion of Aristotle is a joining together of the rational and the irrational?[1]

Never before has it been so possible to confront the perplexing and distressing relationship between mathematics and music. To the medieval mind (which may still be right) the relationship was metaphysical; to the modern mind the metaphysical explanations are too general. In pinning down the precise configurations of the music of mathematics – to be 'music' it must correspond to some aesthetic or spiritual sense hitherto considered deeply irrational – we approach a deeper understanding of ourselves. Such an understanding of brain and mind is scarcely possible through artificial intelligence in its more normal practices, though, of course, that field of enquiry works on related lines. The odd sensation that one has in computer synthesis, of peering into consciousness itself, is an indication to me that music is thereby moving towards the centre of philosophical debate once more. Some thinkers, such as Marvin Minsky, believe that language and linguistics can no longer serve as well as music can as a lever for opening the inner workings of the human mind.

The experience I have described was gleaned at IRCAM in Paris where I spent several months working with MUSIC V[2] and CHANT[3] driven by FORMES[4]. Other systems available there are the Yamaha frequency modulation synthesis system[5] and the 4X real-time digital signal processor[6]. Both of these are well designed to suit a composer's needs: they will produce ready-formed sounds without the composer having to think precisely about every detail, and the intuitive element which a performer brings to music can be directly applied by the composer. The keyboard or faders can, for example, be played *più nobilmente* if that is what is desired. By comparison, the programs which I used require the composer to be precise about every aspect of the sound, to fix exact quantities for every parameter. Because of this using them is a very slow process and doubtless future production at IRCAM will concentrate on the Yamaha and 4X systems; but IRCAM is an institute for research and, in the sense I have described above, the slower and perhaps more demanding systems enhance our understanding in very profound ways. This has not always been fully recognized and needs emphasizing.

One of the problems with using the Yamaha system is that it is difficult to be precise about the frequency modulation system. The relationship between the volume level of the oscillator and the modulation index is a highly complex one: it is measured in terms of the level of the carrier frequency relative to its first- and second-order side bands. For precision in inharmonic modulator–carrier relationships one has to judge by ear, in the dark, or make laborious calculations (a task utterly incommensurate with the extreme ease of manipulating a data entry lever). Computer research, on the other hand, has resulted in the use of the computer to calculate the modulator and carrier necessary to produce the inharmonic spectrum specified.[7] Instead of hit-and-miss attempts to arrive at the desired spectral content, research has enabled the composer to think in precise spectral structures, and to compose consciously in a new and fully lit area where, with the high-level, faster synthesizers, only half-conscious groping previously existed.

MUSIC V was my way of achieving precision. It was particularly good at additive synthesis, and in the IRCAM version allowed such things as the addition of partials to enrich concrete spectra. Recorded sounds could be changed by first analysing the spectrum by FFT (Fast Fourier Transform), then adding in partials at strategic points, for instance so that they beat with existing partials, creating a 'tingle' in a tuned percussion note. The added partial might also make a glissando so that the beating varied in speed over a stretch of time, might get louder, or softer, and so on. Any number of these interactions could be added to transform the original sound in very precise terms. After a period of this sort of work I gained knowledge about the internal life of spectra and the values required to achieve certain sounds. The general nature of such knowledge is quite unobtainable with more musically expressive tools, and with it compositional imagination can attain a new dimension.

Let us take the case of timbre as an example of the interaction of beauty and numbers. Building up sounds from scratch in computer synthesis, one quickly realizes that timbre scarcely exists as a concept. The significant constituents in our perception of timbre are to do with such things as the evolution of the amplitudes of partials in time, the micro-melody played by the fundamental and its fused partials, and above all the mental picture of a wood, metal, gut construct or whatever, which hovers in the mind as a more or less ghostly source of the sound. To change a timbre is to change a melody or a harmony: timbre itself is squeezed out of existence.

To construct a flute timbre one must simulate all the clues the ear seeks to confirm the 'fact' that someone is there blowing a flute: the breathiness, the jumpy micro-melody imposed on the player's vibrato by his or her nervous system, the sound of the lips starting up, and so on. When the 'fact' is convincingly confirmed the mental picture of a flute is born which then unifies many sounds of very different timbre under the heading 'flute'. The listener's perception of these disparate sounds as a flute (and there are more differences between high and low flute notes than between flute and oboe notes) can be seen as an illusion, one of those tricks our order-loving mind plays to reassure us in a world of chaotic flux.

And yet there is a use for the word 'timbre'. Contemporary music is often said to be concerned especially with it. We have all experienced beautiful moments when the play of colour in sound is paramount. We are enchanted and can call it nothing other than a timbral experience. Boulez's orchestration, Stockhausen's electroacoustic manipulations, Ligeti's 'cloud' music, Grisey's or Murail's spectral music – at their best they create a kind of magic we may describe as timbral, though there are many supporting structural parameters as well. They have something else in common too: they are all playing with the identity given to objects by virtue of their having a timbre, in order to create *ambiguity*. The 'timbral experience' is fundamentally one of shifting identities. It occurs when we mistake, however momentarily, one thing for another. The two sounds must, however, be sufficiently distinct in the first place. It is more likely to happen when a certain note on the clarinet is combined with one on the cor anglais, for example, than when a violin is playing with a viola. The fusion of the two spectra into one magical new spectrum does not mean that we necessarily lose sight of the separate instruments (think of Berlioz's moments of timbral mastery) but that there exists a particularly intriguing combination of spectra as well as the original ones.

So we have in timbre a concept that disappears into other things at the computer terminal but which reappears in an indefinable way in aesthetic experience. Since research into acoustics is one of the most exciting obsessions in music at present and probably in the future, this particular interchange between reason and soul is highly illuminating, and brings the 'indefinable' into ever-sharper definition. As Rudolf Steiner believed, man's endeavour should ever be to make things more conscious. That is spiritual development.

My work at IRCAM was preceded by several studies in timbre manipulation. *Inner Light (1)* for seven instruments and tape (1973) uses the tape part to make transitions between spectrum and structural harmony. Simulations of the spectra of the instruments are forever 'opening out' as the partials increase in amplitude until they are equal, and simultaneously sliding in pitch to the nearest harmony note, or 'closing back' into the single-pitch timbre again with the reverse process. The timbre is given greater identity by emerging from the live instrument it is modelled on. The harmony works because the whole piece is based on large harmonic fields with pitches symmetrically arranged over the whole instrumental range, so there are always notes available near to the partials and, at least when the 20 or so harmony notes derived from the 20 partials are joined by instruments playing other pitches from the same pitch field, the structural meaning is easily located by the listener. So timbral structure and harmonic structure form a continuum. 'Cello-ness' is part of the structure, not just an implement to play the structure with.

Inner Light (2) for fifteen instruments, five voices and tape (1977) does similar things using vowel simulations on the tape instead of instrumental simulations. For example, a voice holds a vowel, through a diminuendo. The tape fades in a tetrachord fundamental to the piece with the four pitches placed on the voice's fundamental and on the three formants which I take to characterize each vowel. The tetrachords can mostly be fitted to harmonics above the fundamental. In other words they have an interval content high in 4ths and major 2nds. The tape having faded in an extra emphasis on the current vowel formants, it bows out, cross-fading with four string instruments which then play the remaining two tetrachords of the 'derived set' to complete the 12 pitch classes. Thus the timbre of the voice is an integral part of the serial structure.

In *Inner Light (3)* for orchestra and tape the transitions are from one instrument to another, often accomplished in mid-flight around the concert hall. For instance, a trumpet note is picked out from the orchestral stage, circulates the hall and returns as a clarinet to the stage speakers, whence it is picked up by the live clarinet. The tape part also develops the relationship between timbre and tempo. By the end of the piece the music on tape is both so slow that there is only the reverberating residue of occasional great crashes of noise (slow tempo as timbre) and so fast that the trills of the orchestra are accelerated to a whirring, grainy timbre-texture (fast tempo as timbre).

My aesthetic predilection is to integrate ever more and more; through seamless transitions timbres are integrated with the structure and, as just illustrated, structure is integrated with timbre. The whole structure of the *Inner Light* trilogy is one of expansion, and this occurs very obviously in the third piece where the rapid whirring and the spacious quadraphonic reverberations represent the expanded consciousness of mystical experience, arrived at by a clear speeding up of chord passages of equal note values alternated with (or simultaneous with) a slowing down of others throughout the course of the work. A 'timbral' conclusion to a 'structural' journey.

In several of my works without tape there are passages or movements using interplay between harmony and spectrum. To quote a simple and clear instance, in bars 54–5 of the third movement of *Song Offerings* the soprano sings *senza vibrato* to achieve a more perfect spectral fusion with the two violins and the viola, who play her third, second and fourth partials respectively, and with the clarinet who doubles her. The strings either play in harmonics or *senza vibrato* to hide themselves as fused partials, to become part of the soprano.

During the course of the voice's G the fusion breaks. The string partials come out of hiding and proclaim themselves individuals again with their separate vibratos, and in the same gesture move to harmonically 'correct' locations in the syntax of the piece. The whole movement, whose idea is that of mystical union, constantly moves in and out of fusion and fission. Is an instrumental part only a part or a thing in itself? Such is the ambiguity, the teasing veil of identity.

* * *

When I came to use MUSIC V in *Mortuos Plango, Vivos Voco* for eight-channel tape (1980), my first IRCAM work, it was possible to develop timbral composition.[8] The structure of the piece was entirely based on the spectrum (or timbre) of the large Winchester Cathedral tenor bell. That object pervaded all, and was complex enough, with at least 33 inharmonic partials, to provide ample territory for exploration. Some passages of the piece explored its obscure recesses, yet the ease with which the ear can de-compose or de-fuse the bell spectrum by the end of the piece is simply a result of the fact that the data are not so numerous as to elude the memory. Normally, of course, we don't de-compose spectra; we hear a pitch with timbre. So the aim of the piece might be described as coaxing us to hear abnormally, to hear a spectrum as de-fused individualities. Or rather, the aim is to hear it *both* that way and the normal way, simultaneously.

The spectrum of the bell was analysed with the FFT program at a point half a second after the beginning of the sound: a satisfactorily rich moment neither too distorted by the transient clangour of the attack nor too neutralized by the decay of the upper partials. These latter, present by the thousand in the first micro-seconds, decay very quickly and suggested to me the central image of the piece, the progression from outwardness to inwardness. (The Eastern meditation mantra 'OM' is designed to express the same movement.) The spectacular brilliance of the attack gradually transforms to the prolonged calm of the deep hum note, the last to decay. Doubtless that is why bells are 'sacred' in many cultures.

Another remarkable attribute of great bells is that they contain secondary strike notes which do not show up in any analysis; they are a psycho-acoustic phenomenon. The Winchester Cathedral great tenor bell emits a powerful secondary strike note at 347 Hz (the F above middle C) with a strong element of beating (see Example 1). Its vibrations can be heard from far afield in Winchester to curiously thrilling and disturbing effect. It is a result of the various F harmonic series partials that can be seen in the spectrum (Example 1).

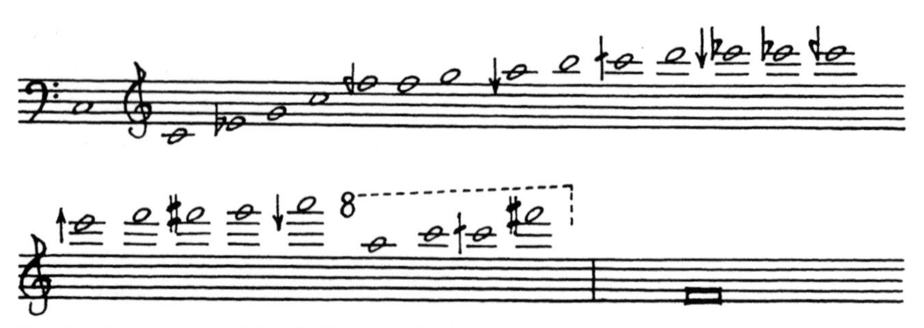

Ex.1: Spectrum of the bell with the secondary strike note

Ex.2: The central pitches of the eight sections of the composition

The eight sections of the work have as their respective central pitches the partials shown in Example 2. The bell spectrum, as shown in Example 1, was synthesized with its appropriate amplitudes. Frequency and amplitude data were then stored in an array available for any sort of mutation. Using MUSIC V I could give the partials any envelope I wished. For instance, I could turn the bell inside out by making the low partials, which normally decay slowly, decay quickly. The normally fast-decaying high partials could be made to decay slowly or even crescendo over selected durations. Modulations from one transposition of the bell to another were achieved by sine-tone glissandi. To avoid over-obvious parallelism, I chose differently located slices of the spectrum as beginning and end sounds, and the current central note was the 'pivot' of the modulation. In this way subsidiary 'bell-tonics' are modulated to, and they are articulated in Schenkerian hierarchies analogous to (but distinct from) the traditional Western tonal system. Each of the eight sections is announced by and based on a bell transposed to the pitches indicated in Example 2, with all its structural implications of secondary pitches.

The straight digitized recording of the Winchester bell in various transpositions was read by the computer in different ways. The IRCAM version of MUSIC V was able to read files forwards or backwards with the option of continuously varying the speed. Often a rapidly oscillating forward and backward reading was made that gave a decrescendo or crescendo of high partials as the attack was left or approached. Rhythmic patterns of great subtlety were easy to devise, by changing the turn-around point, sometimes in interplay with programmed spatial movement. Elsewhere the partials of the bell, or selections from them, were individually distributed around the eight speakers, giving the listener the curious sensation of partials coming from different directions, and even of being inside the bell.

Three separate recordings were made of a boy: chanting the Latin text inscribed on the bell on one partial note, singing all the phonemes of the text separately, and singing a short melody based entirely on the spectrum pitches. I was able to simulate these sounds using the singing synthesis program CHANT developed by Gerald Bennett and Xavier Rodet, though getting the degree of random fluctuation and rudimentary vibrato right for the pure treble voice was difficult at first. In the event, the beginning of the synthetic transformations were often disguised with a 'real' voice fragment. Using another technique, recordings of vowels sung by the boy were digitized. The digitized files were then read by the sound-input modules, looped, and given pitch and amplitude glissandi analogous to those applied to the sinusoidal components in the synthetic bell spectra. Instead of sine tones the boy's synthetic voice sang on the bell partials, and modulations as described previously were effected. Bell-like envelopes were given to some of these bell-sounds-composed-of-boy's-voice. Transformations were also applied to the spectra of the boy's vowels (harmonic spectra), which could be made to slide in pitch and amplitude to the nearest bell equivalents in a bell spectrum (inharmonic spectrum). Such a file could again be read backwards and forwards, giving rapid oscillations of 'boy-ness' with 'bell-ness' in varied rhythms.

* * *

To achieve a frisson of richness it is always necessary to impose rigorous limitations. The fine-tuning of sounds by numbers is a delicate process, and often the adjustments are extremely small: almost, but not quite, imperceptible. One quantity will delight, while another will sound uninteresting. In my next IRCAM work, *Bhakti* for 15 players and quadraphonic tape (1982), I chose to make the tape diverge only slightly from the instrumental sounds of the ensemble.[9] I felt the social and spiritual *differences* between loudspeaker music and live-player music to be aesthetically insupportable and that they should be contradicted – made ambiguous – by having the sound and structure of the tape music as similar as possible to the live music. In addition, the numerous small departures in the tape from strict instrumental imitation can be more effective because the listener gauges so precisely how far removed they are from their 'base'. A foreground can be composed *against* a clear background. Ambiguity is constantly present in that the ear is often unsure whether it is hearing tape or live player. This borderland is intriguing: the numbers necessary to create precise identities and the middle-ground of ambiguity dance in delicate play with the instruments' cultural associations, created by several centuries of usage.

There are 15 instruments for both ensemble and tape (4 woodwind, 3 brass, 5 string, piano, harp and percussion). The family structure was deliberate because many of the sounds on tape are chords made up of one instrument multiplied, thus unifying even further the ensemble's tendency to family chords which consist of *related* instruments (e.g. flute, oboe, clarinet, bass clarinet) rather than groups of the same instrument. In fast taped chordal passages, to give a group of, say, six clarinets the same biting attack, the actual attacks were clipped slightly, and had decay added, thus transforming the sound into something with resonances of other timbres. The sharp attack sounds, such as piano, vibraphone, glockenspiel, harp, crotales, tubular bells and violas pizzicato, remain instantly identifiable. The piano, harp and percussion area of timbre is, again, constantly integrated, not only by adding 'tingling' partials as mentioned earlier, but by hybridization: grafting the attack of one instrument on to the resonance of another; for example, harp/piano, crotale/vibraphone, tubular bell/glockenspiel, harp/crotale, or piano/tubular bell.

There is one point, however (it is a sort of climax or centre), at which the timbral element bursts out of its limitations. The work opens with a sustained blend of disguised timbres on a single note, G, from both instruments and tape. In the ninth movement, this same static G recurs as a tape solo with full timbral exploitation. The idea, of course, is to explore the inner life of a static sound, a spiritual turning-inwards which is at the heart of the notion of *Bhakti*.

The fundamental G is actually sub-audio (1.53121 Hz). In CHANT, which was used for this section, when a fundamental has been set up, its entire series of partials is simulated, but only the ones allowed by a defined dynamic spectral envelope are kept. The envelope is defined by 'formants' with certain bandwidths. When these bandwidths are diminished to approaching zero, the formants become partials and operate as in additive synthesis. One can make them change from harmonics, fused with the fundamental in that way, to formants, fused in a different way. The evolution in time between various degrees of one state or the other can be complex and intriguing. By increasing the excitation time of each formant one progressively adds sustained elements which quickly add up to many thousands of partials.

Only certain of the many thousands of audible partials were used: those emphasizing the pitch class G itself, for example, with beating neighbours at one or two Hz difference. The mass of sound was controlled by vowel-like formants at a higher level which sculpted life-like (but not recognizably vocal) shapes, as did the parameters of vibrato and vocal roughness used by CHANT. Three times this sound transforms in different ways over 60 seconds, reaching saturation point and resolving into silence; the third time distant echoes of it continue to waft around the hall for a further full minute, like dust in outer space.

So the ninth movement of *Bhakti* (there are twelve in all) contains manipulations of sound which give rise to no mental picture. Perception focusses on the changes from fusing partials to moving partials.

* * *

My work at IRCAM since those two pieces has been concerned with completely recognizable sounds and the paradox of their interchangeability. Our memories seem to work in discrete quanta. Our mental pictures are mostly of static images robbed of their element of change. Continuums are slippery, they elude our grasp and puzzle us when thrust under our noses. We try to slice them into comfortable bits.

The completely recognizable sounds I chose to simulate in this new project were: shakuhachi, Indian oboe, koto, temple bell, Tibetan monks and Western plainchant. The simulations were made by Jan Vandenheede, as was the other programming for this project, using CHANT and FORMES.[10] The simulations were extremely delicate and difficult to make. I asked Vandenheede to simulate short characteristic expressive gestures involving transitions from one pitch to another and the expressive ornamentation so characteristic of the behaviour of some of these sounds.[11]

The more deeply one gets to know a sound and its precise numerical formulation, the more individual it seems. The method of synthesis varied greatly between models. The koto, for instance, used formant synthesis mostly, with a little additive synthesis at the attack. The shakuhachi by contrast demanded two levels of formants, one to make the partials and one for their spectral envelope, plus an independent 'effort' system of noise filtering which can be increased as the breathier lower notes are played. The temple bell was made from pure additive synthesis, as precise control is needed over its evolution in time and its vibrating partials.

With such splendid individuals set up it would indeed be paradoxical if they could yet be perceived as part of a continuum as well. We set out to make just that perception possible.

Vandenheede constructed a field using FORMES within which all these instrument models could not only be accommodated but also actually change into each other in smooth, continuous transitions. The difference between this type of work and previous work on timbre transitions is that the latter has always used similar synthesizing methods for all the individual models. That, too, was our initial intention, but we found the resulting identities insufficiently clear, so we adopted the contrasting types of synthesis mentioned above to create clear mental pictures and *then* confronted the problem of their integration. FORMES is a program which organizes the operations of synthesis hierarchically. The hybrid object has any number of 'sons' or voices mostly activated in parallel, which can be drawn upon as required when switches from one voice to another are made, or different ratios of each are present. The evolution of the parameters can be supervised from above, as it were, so that a cross-voice evolution is constantly used as well as within-voice evolution. Vibratos can change into each other, evolutions of pitch in a held note can be interchanged, spectra changed etc. The effect is uncanny and quite different from a fade-in and fade-out mix. This is because the sounds' structures are mutating rather than replacing each other: a much more disturbing effect, because a continuum leads to a mental evolution in the perceiver rather than a mental switch. These changes can be effected three times a second or over 30 seconds: they are easily manipulated (by virtue of complex anterior programming), and this ease is communicated to the listener as readily as is that of a virtuoso violinist darting around the different timbres and effects produced by his or her singular instrument. There is a sense of unity, of everything belonging together.

So one is forced to consider that all these exotic individuals are actually simultaneously one continuum, or their existence flips rapidly and ambiguously between 'is-ness' and 'belonging-ness'. They are indissolubly themselves and yet they belong to a larger entity. The aesthetic urge towards integration without losing individuality is my motive, a motive present from the integration of contrasting subjects in classical sonata form to Stockhausen's "ultimately I want to integrate everything".[12] It is important because art is a means of expanding the tight ego to the larger, more compassionate one, or to the 'egoless-ness' of Buddhism. Art's function is essentially ethical, ultimately spiritual. Any new consciousness born through experiencing, for instance, a timbral transition is a step in this direction, a life changed.

These identities, hybrids and transitions are articulated by melodies. All the models used in the project are essentially monodic. Indeed the oriental cultures to which most of them belong are largely monodic, as is Western plainchant. The sounds are at their most typical when colouring a monodic line, using highly individual ways of moving from note to note. Therefore it is entirely natural that the projected work should be essentially melodic, though in fact I enrich the melodic concept with heterophony and parallelism, and beyond that, with polyphony. 'Texture music', to use the current electro-acoustic cliché, would be entirely out of place with such delicate material. I created a chain of 16 melodies, using partials 6 to 36 of the harmonic series as my intervallic mode, so that the intervals range from small minor 3rd at one end to small quarter-tone at the other. No two intervals are quite the same size, but the difference in feel between a leap of a '6th' from partials 10 to 16 and from 22 to 36 is not because of this slight difference in size but because they are embedded in totally different contexts, in different densities of the mode. I have always been very suspicious of serialism's hypostasis of interval. Most theorists call serialism a system of absolute intervals, recognizable when inverted or, more simply, when they are repeated *regardless of transposition or accompaniment*. The credibility of the system depends on this. But intervals have always been heard in two ways, as interval distance *and* as change from one scale degree to another. I–IV is quite different from III–VI in tonal music. In serialism the scale degrees are not neutralized by atonality, though they are lessened in force, and the 4th does not jump equally from one neutral point to another in all contexts. The points remain obstinately charged and each 4th is different. The abstract and simple relationship necessary for intervallic serialism exists only in theory and the theory is a house built on sand.

Example 3 shows five of the melodies in the chain: every melody embeds the motifs of its neighbours on either side. Each set of motifs (each melody) exists in its own right, and as *part* of two composite melodies; parts and wholes again, this time on the macro-melodic level. The last composite melody of the chain embeds the first melody, so the chain is circular; there is no privileged starting-point or direction. It is a quarry for compositional plunder.

The timbres marked above the melodies clarify the structure. In general they draw attention to the fact that a shape is the same as in a neighbouring melody by preserving the timbre invariant, at least at its purest stage (though it will be often in a state of change, on its way to transforming into a *different* timbre from its partner's timbre in the previous or subsequent related melody). The sense of ambiguity in timbre is present during the transitions, and it is also present in the composite melodies which are both one thing and two things simultaneously (if I have made their internal cohesion as melodic gestalten strong enough). Both (G + H) and (G) and (H) are perceived. It is not particularly easy to perceive a new melody as having parts of an old one in it, so the timbral dimension is used to clarify perception.

We also made canons out of some of these melodies, in which it is easy to perceive the exact structural imitation that is going on. Here we used timbre to hide, not clarify, the structure by having each new entry 'performed' by a different timbre-pair. Ambiguity arises because the ear both hears the same pattern being repeated *and* hears across from one occurrence of the shakuhachi to another, in other words linking the *models* together regardless of difference in melodic motif. With several entries, it becomes hard to tell how many voices there are and the exact relationship between them, but the pure models (not so much the hybrids or transitions) tend to be grasped and linked together.[13] The ear likes simplicity.

The development of this melodic material takes the form of complex polyphony and play with family likenesses. For instance, both (G) and (H) can be played together in time as if they are (G + H), yet they clearly are not. (I) may be added congruently, or (H + I), or both, and (I + J) quickly afterwards in quasi-canon, and so on up to 16 polyphonic voices, or more if the melodies are played in 'parallel' chords. Example 3 shows the richness of imitative texture and a fascinating sense of ambiguity which could result from such polyphony.

The instruments gathered together are all ceremonial or ritualistic; their joining together in one field, one ceremony, has a certain stylistic character. That is of repetitiousness, or hieratic ritual. I believe that a tape solo, using speakers as its only visible performers, is best suited by a certain object-like architectural quality. The expressive evolution and emotional flow of live performance ill-befits loudspeaker music. It must have more of the supra-human passivity of a rock-like object. Thus these melodies, though flowing and fluctuating in the ambiguous mental pictures and timbral teasings they engender, have an exterior of relentless repetitiousness and invariance; fluidity formalized: a dialectic of curve-functions and integers. In such ways do we mirror our indefinable selves.

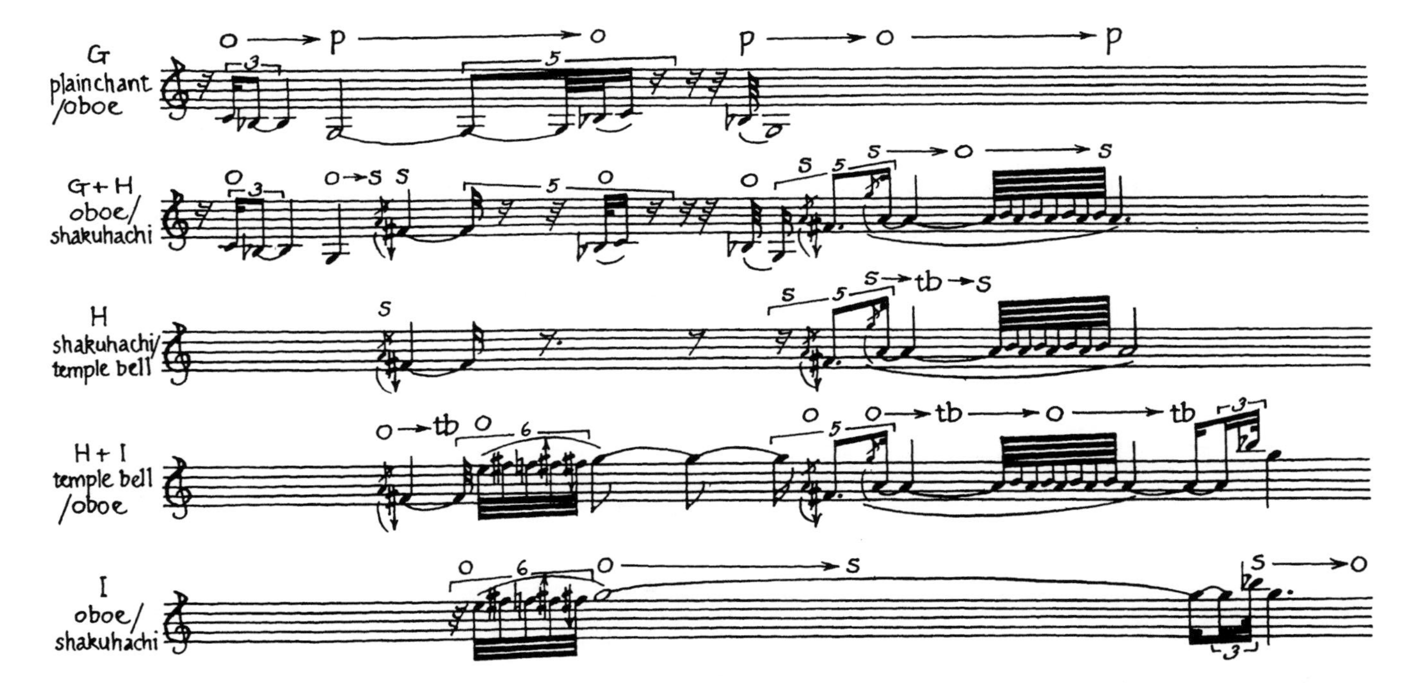

Ex. 3 Five melodies in a chain. A succession of melodies is linked by intervening melodies which add together the melody above and the melody below. Only the beginnings are given here. Each melody moves between two timbres (indicated at the beginning of and above the stave).

Without computers there would be no such mirroring, or at least it would not have the inestimable value of being so conscious a mirroring. One would not become familiar with the dimensions of a new area of sound knowledge, and without knowing dimensions one cannot make those intuitive relationships between man and time (which are yet so precise) of which Stravinsky spoke as being the essence of music. One would have no ruler with which to create beautiful forms, forms so complex that the ruler is forgotten and yet not forgotten. To quote Stravinsky:

> What delivers one from the anguish into which an unrestricted freedom plunges one is the fact that I am always able to turn immediately to concrete things . . . Let me have something finite, definite . . . my freedom will be so much the greater and more meaningful the more narrowly I limit my field of action and the more I surround myself with obstacles.[14]

Computer programs of the type I have been illustrating certainly produce their share of obstacles. But the value of such obstacles is much the same as Stravinsky's: they aid compositional freedom because they 'narrow down' the field of possibilities which then expand to reveal new worlds, just as the microscope selectively narrows down visual perception to reveal new universes. Therefore computers in a unique way *allow* the conception of new areas of work, such as that of the manipulation of what used to be called timbre; that is their primary role in compositional aesthetics. The quest for the new is but an integrated part of the continuum of the old, that being the quest for self-knowledge in which mirrors, like obstacles, have always played their part.

10

A Stubborn Search for Artistic Unity

Tod Machover

My consistent and overriding artistic goal, ever since the earliest compositions of my student days, has been to create musical forms that suggest the search for psychological and spiritual unity beyond seemingly impenetrable veils of confusion and juxtaposition. That such unity exists – *must exist* – is one of my deepest convictions. That modern society presents an unprecedented network of complex parallel worlds, apparently so completely disconnected from each other, only amplifies my feeling that the search for deeper truths lies at the heart of our attempt to keep inter-personal and inter-cultural communication from breaking down and disappearing entirely. Because of its very neutrality and flexibility, computer music technology has long suggested itself to me as the ideal medium in which, paradoxically, the differentiation of musical materials can be intensified while ever-deeper layers of sonic coherence are established.

This article will discuss some aesthetic and technical aspects of this viewpoint, presenting examples from my musical compositions. The most recent of these is *Valis*, an opera based on a novel by Philip K. Dick, on which I am currently working. Besides employing very advanced sonic and visual technology, this work is entirely based on the dramatic treatment of technology as simultaneously unifier, manipulator and destroyer, and is therefore in itself a parable of the aesthetic preoccupations presented here.

Although intended to refer to mathematics, the following statement by

Philip Davis and Reuben Hersh in their book *The Mathematical Experience* could well apply to music:

> Unification, the establishment of a relationship between seemingly diverse objects, is at once one of the great motivating forces and one of the great sources of aesthetic satisfaction . . . The whole object is to create order where previously chaos seemed to reign, to extract structure and invariance from the midst of disarray and turmoil.[1]

It is not by accident that these authors immediately go on to speak of Albert Einstein who perhaps best illustrates the glory – as well as the terrible difficulty – of searching for unity in our century.

It is well known that Einstein, after formulating the two theories of relativity, spent approximately the last thirty years of his life searching for evidence for what he termed a 'Unified Field Theory': one in which "physical reality, space, time, energy, matter are bound together in a single continuum".[2] This idea was clearly against the spirit of the times, for in science, as in art, we live in a period when specialization and differentiation, a tendency to get lost in details, are easier to achieve than synthesis. Einstein simply could not believe that the fragmented view of the world posited by quantum mechanics was more than a passing phase on a path towards more comprehensive knowledge. As he wrote in a much-quoted letter to the physicist Max Born:

> Quantum mechanics is certainly imposing. But an inner voice tells me that it is not yet the real thing. The theory says a lot, but does not really bring us any closer to the secret of the *old one*. I, at any rate, am convinced that He's not playing at dice.[3]

Einstein did not, unfortunately, establish such a Unified Field Theory, and was almost universally ridiculed by his colleagues (although some physicists today believe that such a theory probably is tenable and that its proof is imminent). But the courage and persistence of his search is perhaps the deepest metaphor for our age: it expresses both the powerful intuitive knowledge of the existence of such unity, *and* the extent to which our society has lost track of such goals in a morass of superficial details.

The drama lies in the fact that we have reached intellectual awareness of this complex world but no coherent and unified view of human activity. What are the human truths that underlie cultural specificities? What mental and emotional processes give a common link to our separate sensory experiences? What basic concepts could help us integrate the greatest diversity into a rich and coherent vision?

We must find a way to contemplate this complex world as a whole, free from personal moment-to-moment existence dominated by ever-changing perceptions, wishes, hopes, and primitive feelings. Again, it was Einstein who said that only in such an image could one hope

> to place the center of gravity of emotional life, in order to attain the peace and security that one cannot find within the narrow confines of swirling personal experience.[4]

Yet in order to strive for unity we must search, from far without and from deep within, to view objectively enough to perceive the disappearance of differences; we must part the layers of illusion to arrive at seeds of fundamental truth. The first method could be exemplified by our culture's exploration of space and by the acceptance of the existence of a larger universe, the essence of which has long been understood and was powerfully expressed centuries ago by Chaucer at the end of *Troilus and Criseyde*:

> And whan that he [Troilus] was slayn in this manere,
> His lighte goost ful blisfully is went
> Up to the holownesse of the seventh sphere,
> In convers letinge very element;
> And ther he saugh, with ful avysement,
> The erratik sterres, herkininge armonye
> With sownes fulle of hevinish melodye.
>
> And doun from thennes faste he gan avyse
> This litel spot of erthe, that with the see
> Embraced is, and fully gan despyse
> This wrecched world, and held al vanitee
> To respect of the pleyn felicitee
> That is in hevene above; and at the laste,
> Ther he was slayn, his looking down he caste;
>
> And in him-self he lough right at the wo
> Of hem that wepten for his deeth so faste;
> And dampned al our werk that folweth so
> The blinde lust, the which that may not laste,
> And sholden al our herte on hevene caste . . .[5]

Such is an extreme view of the unifying and consoling aspect of a large, extra-personal, all-encompassing world vision. But the opposite approach is at least as important: that of delving *beneath* the surface of physical reality or human behaviour to find ever more basic functional principles. This was the goal of all atomic physics and the importance of Freud's work, and is currently the study of cognitive science which, by drawing on many different disciplines, attempts to analyse and explain the human thinking mechanism.[6]

Art is also a powerful mechanism for exploring ourselves, our relationship with others and to the outside world. Art is not meant to instruct, nor to provide a one-sided theoretical picture of reality. Rather it should allow us to *experience directly*, through abstract formal processes, a rich interconnection of human emotions and thoughts and – by seeing them objectively – to understand them better. Music is therefore not merely metaphor, but is capable of conveying ideas of truth. This view is the subject of an entire novel: Kim Stanley Robinson's *The Memory of Whiteness*.[7] The book centres on Arthur Holywelkin, a physicist living about 1000 years in the future, whose physical theories, which have superseded Einstein's, are conveyed through music and through the ultimate musical instrument which he invents for the purpose. Robinson's solar system contains a rich mixture of cultures, based on divergent notions of political order, but unified by an appreciation of music as a language that speaks, without need of translation, to something deep in human nature. "Music was the clearest and most powerful analogy because it came closest to pure idea . . . The universe was a music of ideas. And he could write that music".[8]

This rather extreme view is close to my own. Art that is merely self-referential, that speaks only of its own materials and forms, is sterile and superficial. The elements of artistic language must be carefully adapted by the artist to represent a particular human vision. It is precisely this delicate marriage of deep, original vision with a fitting, concise and expressive form that ultimately determines the value of a work of art. Art *can* and *must* represent human reality, but needs to objectify and elevate this reality by giving it form. Music allows us both to observe reality by understanding the behaviour of musical materials in ever-deeper ways, and to express a view of reality through the use of these materials.

The conviction to which I keep returning is that there exists a degree of unity far deeper than that which is suggested by the surface differences and complexities surrounding us. I long to compose music that conveys clearly this sense of equilibrium and balance. But I have not yet found such perfect unity. Therefore, my music expresses a *search*, sometimes striving and impatient, once in a while more calm and welcoming. The materials that I use for my work and the forms that give them life are chosen because they seem perfectly suited to this central vision.

But there exists a powerful, tension-generating paradox in my work. My goal is to achieve unity through *inclusion* rather than *exclusion*. I look to find a place for so many different sounds and melodies and textures and harmonies that each piece hovers on the brink of exploding from too much contrast, too many differences. In this way, I am perhaps a spiritual descendant of such American composers as Charles Ives and Elliott Carter. I will only be satisfied when a place is found for everything in the world that stimulates me, everything that I love; and it is out of this chaos, this sonic jungle, that I attempt to forge unity.

I am not interested in such restrictive artistic movements as serial, minimal, neo-expressionist, new wave, or 'sound art'. They all simplify the world by choosing a safe and limited corner of it. I am motivated rather by the ideal of Bach, who knew how to find a place for every musical idea that his epoch had discovered. And he knew how to forge a language, a context, in which all of these elements found a natural place. It is surely of this all-encompassing vision that Beethoven thought when he punned on Bach's name: "Nicht Bach; Meer sollte es heissen" ("Not Brook; his name should be Ocean"). I am not lucky enough to possess such a comprehensive and total vision. But my fundamental task is to get as close to one as possible, to try as hard as I can to incorporate in this tapestry every musical element that I find stimulating, and to document this journey in my compositions.

The personal language that I have forged attempts to satisfy these conditions. I have looked for a set of basic sonic materials, syntactic constructions, expressive states and formal principles that allow me diversity and control simultaneously.

The main parameters of any musical language are melody, harmony, rhythm and timbre. I attempt to enrich each of these parameters as much as I can. But there are limits to our perception and, specifically, our auditory understanding. If all four elements are simultaneously at a maximum level of intense activity, we perceive only confusion. For instance, it would be easier to perceive the rhythmic distinctness of four musical lines moving at different speeds (a polyrhythm, say, of 4 : 5 : 6 : 7) with complex accented patterns if the general harmonic rhythm is slow, rather than if chords are whizzing by at the speed of the shortest note value. The experience of serial music taught us how quickly one loses track of complexity when it is present at the same time on all musical levels.

A new degree of sensitivity is required for the intelligent use of the vast timbral resources which have been discovered since the beginning of this century, and are being so rapidly expanded through electronic means. Once a mere decoration of the other musical elements (unlike the prominence long given to this parameter in non-Western, and especially East Asian, music), the way sound colour changes over time has become a significant tool by which musical meaning can be conveyed. But, here too, the complexity principle reigns – delicate transformations of complex inharmonic spectra are perceived most clearly when harmony is virtually non-existent, melody very simple, and rhythm completely static.

The trick for today's composer is to balance or 'unify' these four elements, to make sure that they complement each other. It is as if the four parameters were four passengers in a rather unstable rowing boat, each seated in one of the four corners. With all four sitting still the boat remains stable and well balanced; when one moves, stands up, or jumps up quickly the others must stay more still than ever, moving only to redistribute weight and avoid capsizing, often leaning over the edge of the boat to do so. Imagine how extreme the situation would become if one of the boaters were to dive overboard for a swim!

The composer needs to manipulate and control these parameters to achieve various points on the continuum between independence and interdependence. Diversity and complexity are quickly heard as the activity of each parameter is increased. Unity is achieved in two ways. In the first, one of the parameters is clearly established as being more prominent than the others, which then take the role of embellishers. I have often used this method to allow, for instance, a rich and florid melodic line to express the continuity of a particular musical section (*Nature's Breath*), or alternately harmony (first movement of *String Quartet no.1*), rhythm (last movement of the same piece) or timbre (*Fusione Fugace*). In each of these cases, one parameter is given structural importance and defines the musical shape for a particular span of time.

The second method for achieving multi-parametric unity is to moderate the activity of *all* parameters, so that none is too dense and all shed light on the others. Since my desire is to create states of imbalance, I tend to use this form of writing as a special case, often appearing near the end of compositions, when the necessary musical elements have had a chance to cohere.

Computer technology represents an enormously powerful tool for the manipulation and definition of such musical materials. In the last few years I have written several texts[9] which describe in some detail the technical procedures that I have developed for computer music, so I prefer to devote myself here to larger musical issues. The computer can, in fact, be used to heighten both ends of the expressive spectrum which I have described, by emphasizing the *differences* between musical elements, and by drawing strong perceptible *connections* between them. It can operate effectively on all levels of musical discourse: on sound-objects, musical phrases and the overall formal scheme. It has become an indispensable tool to me in the forging of my musical language.

The first important feature of computer music technology is its *neutrality*: the fact that, in theory at least, any sound structure can be produced or manipulated on the computer. In terms of sounds themselves, this means that new instruments can be created which bear no relationship to the historical references that we associate with traditional instruments. To help achieve the extreme contrasts that I look for in my music (as the necessary starting point for my dramatic formal schemes), the computer permits a wide range of such sounds to be created, either by modifying the behaviour or sound of existing instruments, or by creating instruments that do not exist at all in the physical world. In this way, the 'players' in the musical drama can have a range of diverse characteristics inconceivable in the purely instrumental sphere.

The ability of the computer to increase diversity is counterbalanced by the principle of *continuous transformation* which it also makes possible. The liberation of timbre from a supportive role to a fundamental structural level is one of the most important features of twentieth-century music. The computer has provided the ideal tool for constructing new timbres (and for transforming those that exist already) from constituent elements, harmonics or partials that can be controlled independently and precisely. In non-electronic music, the musical instrument is the smallest sonic unit. With computer music, the least common sonic denominator is not an individual sound-object, but the constituent sonic partials that make up that sound. It is therefore possible to use these partials to change gradually from one sound-object to another or simply to configure or 'fuse' a sound-object out of an indistinct timbral background.

This possibility is of great importance for two fundamental reasons: first, a connection can then be established between instruments which can be given musical sense as a unification of opposites, in line with my dramatic goals;

second, a listener can be made aware of the process of transformation from one sonic object to another as a primary musical event, and can therefore perceive a path of change rather than a goal. Such a phenomenon is without precedent in Western music.

The use of fine control of individual partials to integrate separate sound-objects can also be applied to the interior unification of timbre itself. A separate partial can be thought of as a pitch: a frequency component with no timbral dimension. As several partials are added together, a chord of harmony is perceived, especially if these partials are in an harmonic, or 'tempered', relationship to each other. If this process is continued and either a threshold level of more than a certain number of partials (probably around 20), or a particularly irregular and non-harmonic grouping of these partials, is reached, the sound no longer resembles a chord, but rather a colour or timbre, with brightness and density characteristics but a low level of transparency and few individual partials. When the spectrum is made even more complex, either by saturating certain frequency regions or by distributing multiple partials throughout the frequency range, this impression of timbre turns to noise; sounds too complex to analyse easily. To close the transformational cycle, a noise which reaches a maximum level of density, and which in fact completely saturates the frequency space, is again condensed by our perceptual system and begins resonating at a frequency or pitch somewhere around the middle of the saturated spectrum. The most complex sound is perceived as being identical with the most simple. The importance of this concept cannot be overstated: the computer control of individual partials allows the four different timbral states to be unified in one transformational continuum and no longer to be regarded as four separate, vaguely related musical elements; they are different states of the same acoustic–aesthetic phenomenon. Just as the computer allows the differences between these states to be explored, articulated and intensified, so it makes it possible for them to be linked together in an organic, and perceptually comprehensible, discourse.

This same principle can apply on a larger formal scale when the computer is used to generate a *continuous structural background* for diverse compositional elements. Since the computer can interpolate and control transformations with great finesse, it can easily reproduce common elements from a variety of musical events and turn these into a sort of obligato or wallpaper. This can be done with timbre, for example, if an analysis is done of the most prominent instruments and the resultant data are frozen in time or made to evolve very slowly from one timbral state to another. When the timbral transition is given enough definition of fine detail, and when the compositional connection is made between this level and the sounds of individual instruments, the effect can be to provide overall cohesion. The same procedure can be applied to motivic or harmonic material played by separate instruments and repeated or held in time by the computer. I often conceive of this process in the opposite way: it is possible to generate computer music material of such richness and complexity that many individual musical gestures can be given definition and provided with a meaningful context because they draw from these 'background' resources. The relationship is much like that between the functional and directional quality of tonal chord progressions, and the light that they shed on melodic lines; the two may, in certain cases, be quite independent of each other.

A final category of operations that is specific to the computer is the *conceptual description* of musical processes. Computer programs, especially those based on artificial intelligence concepts, allow for processes to be described that are independent of the materials on which they operate. A pattern of transformation can be encoded, as could an algorithm of permutation, that would change or influence various musical elements. It is possible to manipulate these processes so that the listener becomes aware of a certain pattern of change, rather than of the particular sounds that are being influenced. In this way, the composer can unify quite diverse musical elements by applying the same or similar structural transformations to them, another powerful way of creating unity out of dissimilar material.

Much as computer control over timbre has allowed composers to create a totally new listening experience, the computer manipulation of formal structures can result in a fundamentally new musical experience. Delicate patterns of change may be followed consciously as the listener tries to grasp the inner puzzle from which the particular composition was generated. Through object-oriented programming, structure will become a musical parameter as important and as consciously perceptible (while also appealing to subconscious processes) as pitch, harmony, timbre and rhythm.

Music conveys ideas by causing sonic material to change over time. We learn and understand musical messages by observing these changes, by comparing differences, by seeing the same material in new contexts, by watching two ideas become one. Form is the most important musical element, for through form we manipulate musical ideas and create magical transformations.

* * *

There are many formal principles that can be used to create tangible tension between unity and diversity. In each of my pieces involving computer electronics I have used the above-mentioned computer music principles to dramatize ideas about chaos and unity. The formal principles that these different works utilize demonstrate how much my thinking has evolved in this area since 1978.

In *Light* (1979)[10] for chamber ensemble and computer-generated sounds, my goal was to confront the listener with extensive juxtaposition of contrasting elements in all layers of the musical texture. With each musical element originating from common harmonic and melodic cells, the formal structure became a gradual stripping away of the layers of difference until their similarity, or unity, was perceptible. On the simplest level, the musical forces are divided into three groups: an ensemble of 14 instrumentalists, and two separate four-track digital tapes diffused through two sets of loudspeakers, completely separated from each other in the listening space. The first tape uses sonorities and musical materials with a strong resemblance to the world of the live instrumentalists, while the second plays music which is much more specific to computer synthesis, generally slowly evolving time-variant inharmonic spectra.

The instrumental ensemble functions on a level somewhere between the two tapes. It is divided into four sub-groups, each of which develops a distinct set of musical tendencies (melodic, harmonic, timbral and rhythmic). The two tapes reflect two quite opposed compositional attitudes. Tape 1 exaggerates and intensifies characteristics of the traditional instruments, becoming a sort of 'super orchestra' by playing more densely, faster, and with more rapidly shifting timbres than would be possible with the 'real' instruments. On the surface, Tape 2 seems to bear little resemblance to the instrumental material, but in reality its complex timbres are all based on transformed spectral analyses of these same instruments. Yet, rather than inviting a rapid analytic parsing of separate events, the complex timbres of Tape 2 suggest a completely different listening attitude: confronted by a complex sonic phenomenon the listener becomes conscious of its constituent elements over the course of the composition and, in a sense, is invited inside the sound. Rather than following a developmental narrative, one becomes aware of a reality that was present all along. The computer materials were used to heighten the contrasts and juxtapositions in *Light*, well beyond what would have been possible with traditional instruments alone.

The piece begins by emphasizing the distinctness of all its various layers.

Each group follows its own developmental principles in a section that culminates in a series of cadenzas. After each group has had its say, all material is combined in the large solo of Tape 1 which builds until the final crashes of Tape 2. In the quiet that follows, a new, more homogeneous, order is built up gradually, and leads to a final section of delicate chamber music, where equality prevails among all the diverse elements.

Over the course of the piece, the timbral transitions as well as the references drawn gradually between Tape 2 and the instrumental ensemble, serve to underline the common link between all the elements. In *Light* the computer was used to help create a sort of wild polyphony, where many musical events sounded simultaneously, almost to the point of cacophony, but where each one shed light on the others. The computer intensified contrast, and facilitated the task of mediation.

My approach in *Spectres Parisiens* for flute, french horn, cello, chamber orchestra and computer tape (1984)[11] was quite different. Rather than trying to create a dense world of simultaneously differentiated musical events, I wanted to establish a linear journey where many diverse states could be unified over time. At any one moment in this piece, all the instrumental and electronic forces contribute to a common texture and musical context. The connecting or unifying element is long-range spectral transition, the sort of background transformation that I described above. The entire structure of the piece is defined by a dramatic passage through the pitch–harmony–timbre–noise continuum discussed earlier in this chapter.

The computer part furnishes a very slow timbral transition and defines the formal movement and continuity of the piece. This is the 'ocean' in which the individual notes, phrases and sections find their mediator. There is, in fact, a rather striking contrast between the widely differentiated playing styles and expressive attitude of the instrumentalists, and the more objective, unperturbed overview heard in the electronics. The solo instruments give, in a sense, specific instances of each timbral state and are each prominent at one time or another. The orchestra mediates between instruments and electronics, alternating between fused ensemble textures (approaching the dense spectra of the tape) and an intricate polyphony (mirroring or answering the soloists).

Each solo instrument provided the model for a complex spectrum which was generated on the computer to give the sense of background continuity described above. In generating the flute noises, the fine distinction made by a flautist between pure breath, breathy sound, complex sound (multiphonic) and pure pitch was emphasized. The flute was used as the basis for a noise pseudo-spectrum that transforms gradually into delicate inharmonics. After experimenting a great deal with the french horn, I became especially attracted by the complex sonorities that can be produced by singing and playing at the same time. The singing can enter and disappear subtly during a note, and the effect produced ranges from gentle beating, through reinforced partials to complex inharmonic spectra. I have always been struck by the 'metallic' character of most inharmonic sound, and thus chose the horn as the model for gradually evolving inharmonic spectra. With the cello, the presence of the harmonic series is quite audible, as the slightest fluctuations of bow pressure and speed break down the simple fusion of a note and bring to prominence any number of partials. The cello was thus used as the model for an elaboration of a 'fake' harmonic series.

Spectres Parisiens is a further step beyond *Light* along the evolutionary path towards unity. Whereas in the earlier piece correspondences were shown between different levels of musical material on a sporadic basis, as bits of sunlight shine momentarily through passing clouds, the later piece feels cohesive both in formal momentum and in background texture. It is as if the transitional materials lurking between and inside separate events have taken on greater importance although a distinct difference is still felt between the timbral world of the computer tape and the expressive linear playing of the instrumental ensemble. I think that I have subconsciously expressed in *Spectres Parisiens* a desire, or *will*, to unite the timbral language developed during my years in Paris (1978–85) with the harmonic and melodic language forged in New York (1973–8). An almost unbearable tension is felt, since these two worlds often clash and do not sit comfortably together – not yet, at least. This feeling of forcing, of trying consciously to forge a unity, adds greatly to the expressive power of the work.

* * *

The next distinct step along the path – in search of unity at many different levels – is my latest piece, *Valis*. I feel somewhat uneasy about discussing a composition which is still being written, but will make an exception here because the issues which I am treating are so central to the thesis of the chapter, and because the musical considerations found in *Valis* will permit me to make some comments about computer music concepts which I am developing at the present time.

Valis was one of the last books written by Philip K. Dick.[12] The Californian author, who died in 1982, was mainly known for his science fiction works, although from the beginning of his career he was clearly preoccupied with very fundamental human issues, often only loosely connected with the science fiction genre. A connecting theme in his work is the fact that the world we see around us is not necessarily the only world that exists, and is perhaps not even the 'true' world. His texts, whether they use metaphors of space travel, advanced technology, mind-altering drugs, or metaphysical religion, present complex labyrinths so that the reader is never sure where one world begins and another ends, nor how these various worlds might coexist. Dick's greatness lies in the fact that he managed to remain true to his belief that a connection between these multiple realities *does* in fact exist, but refrained from giving a simplistic resolution to this cosmic puzzle in any of his best books. His work has appealed to me so strongly because, as I expressed earlier, Dick writes about the *search* for metaphysical unity, rather than of its sure, and complacent, existence.

'Valis' (Very Active Living Intelligence System) is Dick's most potent representation of what such a truth might be. The opera I am shaping from his novel dramatizes a Parsifal-like search for this truth and the elusiveness, frustration and even madness that the path brings in its wake.

Valis, which has been commissioned by IRCAM for the tenth anniversary of the Centre Georges Pompidou in Paris, will be an almost totally 'electronic' opera. It will have three final forms: a live show designed for the Forum of the Centre, an enormous hole in the entrance hall which allows spectators to view the opera from traditional seating or from observation posts high above the scenic level; a public installation which will remain in the Forum for two months, and which will allow the public to circulate among the video projections and sound sources of the computer-circuit-like performance space; and a video disc version, meant to be viewed on television. The central character is an actor whose voice will be transformed and changed into music by a live computer system. There will be four singers whose voices are also treated by this system, two instrumentalists (one keyboard, the other percussion), and an array of live and pre-recorded computer-generated and -transformed sound. The entire sound world of the opera will be constructed from the human voice, computerized extensions of the voice, and complex inharmonic sounds based on these extensions. These limited materials will allow me great diversity, and a rich integration, both of which are necessary for my dramatic purposes. The same concept is mirrored in the highly complex video installation for the work, which will use very large single-screen projections and a system of multiple monitors distributed irregularly in the performance space, allowing simultaneous independent images or the distribution of a large image across distinct and separate points. As with the music, the images will unite projections of the live performers, transformations of these images, and completely synthetic, computer-generated images.

Valis is the only autobiographical book that Dick wrote. It starts out as a narrative about a character named Horselover Fat who, we learn after a short while, is the double, the split personality, of Dick himself (Philip from Greek meaning 'lover of horses'; Dick from German meaning 'fat'). Dick's personality has been split by life's traumas into Fat, who is the indomitably cheerful, if nutty, searcher for metaphysical explanation – the science fiction writer perhaps – and Phil, who is the rational and somewhat weary documenter of mundane human existence. In many ways, the entire drama is the search to unify these two personalities.

The opera begins with a representation of a bizarre experience that Fat has had sometime before the beginning of the work's actual narrative. He has been invaded or pierced by a beam of pink light which enters his brain one day and changes his perception of the world: time stops and many historical moments, from ancient Rome to the world of the far future, are superimposed or cross-faded; he becomes aware of much knowledge that he didn't have before, such as the ancient Greek language and the fact that his son must be taken immediately to a doctor to avoid a terminal illness; he hallucinates images of most visual art of the twentieth century, sent to him at enormous speeds, yet all comprehensible; he is given an idea of the meaning of the world, and of a force, 'Valis', which is behind everything.

Fat returns to daily life, much shaken by this experience. The first part of the opera presents his life on different levels. We learn that a particular event in his life, the attempt to save a girlfriend, has pushed him to the point of strain and mental depression where the bizarre pink light experience would be possible. One level of the action, mostly seen on video monitors as indistinct, timeless memories, involves flashes from the scenes with this woman, Gloria, where Fat tries unsuccessfully to save her.

Much of the action which we observe on the stage involves Fat's day-to-day life. He is in the midst of another hopeless love affair, tries to explain his pink light experience to a group of friends (among them his double, Phil), and becomes so upset that he ends up in a mental hospital for a short but intensive stay.

As we observe the deterioration of Fat's daily life, we become aware of the fact that he is spending all his nights writing his 'exigesis': an attempt to make theoretical sense of the pink light experience, and a description of what he considers to be fundamental reality. A characteristic passage is:

> One Mind there is; but under it two principles contend. The Mind lets in the light, then the dark, in interaction; so time is generated . . . He lived a long time ago, but he is still alive . . . I term the Immortal one a *plasmate*, because it is a form of energy; it is living information. From loss and grief the Mind has become deranged.[13]

The writings are an odd combination of mysticism, high technology and complete paranoia, a desperate attempt to make sense of, to 'unify', an experience that is beyond his grasp, although really a string of disconnected theories, each one pretending to be definitive. As the narrator Phil says of Fat's search:

> Over a long period of time, Fat developed a lot of unusual theories to account for his experience. One in particular struck me as interesting. This theory held that in actuality he wasn't experiencing anything at all. Sites of his brain were being selectively stimulated by light energy beams emanating from far off. These selective brain-site stimulations generated in his head the *impression* – for him – that he was in fact seeing and hearing words, pictures, figures of people, printed pages, in short, God and God's message. But he really only imagined he experienced these things. What struck me was the oddity of a lunatic discounting his hallucinations in this sophisticated manner. Fat had intellectually dealt himself out of the game of madness while still enjoying its sights and sounds . . . He stayed up to four a.m. every night scratching away in his journal. I suppose all the secrets of the universe lay in it somewhere amid the rubble.[14]

During the first half of the opera, these planes of Fat's existence increasingly disconnect from each other. His love life falls to pieces, his journal entries become wilder and more fantastical, and finally he is on the verge of total mental collapse. His friends, of whom Phil is one, in an attempt to distract him from his problems, offer to take him to see *Valis*, a new science fiction film. The film is the centrepiece of the opera, and the dividing line between the two symmetrical halves. Instead of being an innocuous science fiction story, *Valis* turns out to be a kaleidoscopic and surreal portrayal of the pink light vision and various hallucinations that Fat has experienced. Turned into narrative form and objectified for the first time, it bolsters Fat's belief in the reality of his experiences, and his friends awaken to the fact that Fat may not be as crazy as he seems! The second half of the opera becomes a search for the people who made the *Valis* film, to discover who they are, why they made it, and what truth lies within its message.

It turns out that *Valis* has been made by two rock musicians, Eric and Linda Lampton, who live in northern California. Fat and friends arrange to meet them, and are confronted with rather bizarre characters who seem as much tricksters (and slightly mad themselves) as they do saints. However, they begin to explain the ideas behind the film, all of which point to the existence of a superhuman unifying force with which they are in touch.

However, when pressed to give too many details, the Lamptons admit that it is not they who have the direct knowledge of 'Valis' – rather it is Mini, a dwarfed computer music composer who specializes in what he calls 'synchronicity music'. This music, generated by a large complex system which Mini has invented, is capable of transmitting direct knowledge through sound. In fact, it seems that Mini has written the music for the *Valis* film, and has used the soundtrack to transmit all the film's messages subliminally. To explain what the idea of 'Valis' is, Mini gives a complex and powerful performance on his system.

After hearing this music, Fat and his friends are informed that the real truth behind 'Valis' lies one step further. The Lamptons have a two-year-old daughter, Sophia, who has been sent to earth by 'Valis', and who possesses all of the truth in the universe. They are introduced to this girl, who seems to be part real, part hologram. She delivers a gentle and comforting speech, which reassures the searchers, and is very humanistic and down-to-earth in its message. This meeting is the climax and heart of the action. All seems to be explained, and for the first time, Horselover Fat disappears. The split in Phil's psyche has been repaired and his personality has been unified, and he says/sings:

> It was as if I had been shaking all my life, from a chronic undercurrent of fear. Now the fear had died. I could forget the dead girl. The universe itself, on its macrocosmic level, could now cease to grieve. The wound had healed.[15]

However, the action has one final turn. As Sophia is presenting her truth, we become aware that Mini, who is still seated behind his computerized mixing console, is in fact manipulating Sophia; she is, in some unclear way, a synthetic and unreal product of the technologist's laboratory. Mini increases the intensity of Sophia's performance more and more, moving far from the initial balance and calm, and finally overloads the system, causing Sophia to explode before our eyes in a pink light which ironically and painfully brings the action full circle to the opening vision.

The coda sees Horselover Fat return to join Phil Dick. The dream of unity has been broken and, again, the reaction of the two alter-egos is different:

> Fat: Sometimes I dream . . .
> Phil: I'll put that on your gravestone.
> Fat: You rob me of hope.
> Phil: I rob you of nothing because there is nothing.
> Fat: You don't think that I should look for him?
> Phil: Where the hell are you going to look?
> Fat: I don't know. But I won't give up. I never will. He is somewhere. I know it.
> Phil, as narrator (voice off): I have a sense of the goodness of man, these days. I don't know where this sense comes from, but I feel it. . . I ask myself, Is Fat having another experience? Is the beam of pink light back, firing new and vaster information to him? Is it narrowing his search down? . . . My search kept me at home; I sat before the TV set in my living room. I sat: I waited: I watched. I kept myself awake. As we had been told, originally, long ago, to do; kept my commission.[16]

In order to convey the substance of this complex drama, I have searched for a musical language capable of even greater diversity, yet stronger unity, than any I have used in the past. The computer is the essential tool in this language, and the project would have been inconceivable without it.

The musical treatment of the two halves of the drama differs greatly. Almost all of the sound material in the first half is taken from the human voice. The pink light explosion at the beginning contains fast-cut and overlaid snippets of the material for the entire piece. It will be mixed in such a way that although the overall effect will be extremely intense, it will still be possible for the listener to be aware of, and to remember, something of its constituent elements. On one level, the entire sonic process of the opera will be the transformation from the initial incomprehension of the opening explosion to the gradual introduction and reassociation of the constituent elements to build up the second explosion at Sophia's death. A false sense of unity leads to curiosity and finally to understanding, which is, however, immediately thwarted and denied.

The three levels of action for the first half will be treated in contrasting ways, all revolving around the human voice. The Gloria story, which is abstract and almost subliminal (but holds the key to Fat's mental instability and ends the first half with a representation of Gloria's suicide), will use pre-recorded and treated vocal sounds. These sounds will be on the verge of unintelligibility, and the computer will be used to create rich, dense spectra based on consonants. Words from Gloria will be stretched out in time, spectrally enhanced, and combined with purely synthetic noise sources. The effect will be of semi-articulated vocal noises emerging and fusing into audible objects from a slow-changing noise-like synthetic background. This music will play delicately on the frontier between noise and word.

The opposite vocal treatment will be found in the imaginary world of Fat's exegesis, which will also exist mostly on video and sound tape. Fat will be seen dreaming, writing, amidst overlapping scenes from his journal, mixing timeless landscapes, icons, futuristic technology and metaphysical imagery. Here the musical material will be a combination of pre-recorded chanting, slightly melismatic but rather simple (drawn from the exigesis texts), and complex time-variant inharmonic spectra based on vowel sounds. Individual sung notes will seem to fuse out of a wallpaper of inharmonic spectra; fluid timbres will emerge out of a simple melodic line. There will be a constant ambiguity between clear, sung voices, and a redistribution of the spectral components that make up these voices. This music will walk the fine line between pitch and timbre.

The everyday world of Horselover Fat, and the Fat character in particular, will mediate between these two extremes. Here, however, the common element will be the spoken word. I have chosen an actor, rather than a singer, to play Horselover Fat. His speech will begin rather mundanely – clear and intelligible. Over the course of the first half of the opera, a live digital transformation system (4X machine) will begin to change this speech. At first, the transformations will simply be an enhancement of speech, adding exaggerated noise components or inharmonic shadowing to suggest an almost imperceptible relationship with the other strata of musical activity. But then the transformation process will become more and more complex. Fat's words will go through the computer's 'black box' and become integrated with a variety of different processes. A single word may be simultaneously broken up, edited into several components, shifted in pitch repeatedly as well as stretched out in time, combined with a spectrally modified resynthesis of the same word, and the whole given a unified amplitude modulation. In a first stage, the word would become a seed for an instantaneously generated computer sound poem, shaped in such a way as to colour the particular word used. Each delicate and complex sound-object will voyage rapidly (each one might last no longer than a few richly detailed seconds) between the extremes of timbre and noise, while keeping textual intelligibility. Gradually, the transformations will become oriented less towards timbre and more towards pitch, to the extent that they actually become singing voices. At this stage, words from Fat's mouth will go through the computer system and come out the other end as song!

These various processes culminate in the finale of the first half. The inarticulate sounds accompanying the Gloria story become thunderous and terrifyingly grating; the exegesis music has become completely disembodied from the text and turned into wildly fluctuating inharmonic partials, merging into vocal sounds for only brief instants; Fat's friends talk and sing alternately, commenting on the various events. These layers of music become so dense and varied that they reach a point of stasis, a principle that I have often used in my work: all is so complex that the ear simplifies and hears it as a more fundamental unit on a higher level of perception. The swirling sounds are heard as a static chord which has a thick texture, but no independent interior parts. In the middle of this swirling halo, Phil Dick delivers the monologue that ends the first half. He almost whispers a very moving passage about personal pain, loss and disillusionment:

> I have had dreams of another place myself. In the dreams I felt deep, comfortable and familiar love toward my wife, the kind of love which grows only with the passage of many years. But how do I even know that, since I have never had anyone to feel such love for? Are we all like Horselover Fat, but don't know it? How many worlds do we exist in simultaneously?[17]

His simple words blossom into mellifluous origami flowers as they undergo live computer treatment.

Up to this point, all the opera's sound material (ranging from noise to inharmonic timbre to speech to singing) has been modelled on the human voice. The computer will have been used to enhance the differences of each of these sonic worlds, as well as to link them through intermediary states of transition.

When the *Valis* film is reached at the opera's centre, a change takes place. The new unity arrived at by the over-saturation of voice-oriented events turns into pure music. This pattern of congealing into a single entity is emphasized by the fact that the multiple planes of dramatic action represented on three levels throughout the first half (live action, multiple and fragmented visual representation on many monitors etc.) are focussed for the first time on one large-screen projection – a single event to perceive. The 'ease' of following this unitary presentation is counterbalanced by the fact that the action of the film (which lasts about eight minutes) is itself highly fragmented and difficult to follow. For the first time in the work, however, a computer-generated soundtrack is heard which is independent of the vocal text or singing. The film serves to mark the important progression from speech (through noise and timbre) to true song. It also resumes the main themes and images of the first half and, by introducing the Lamptons and Mini, provides the impetus for the search that leads to the end.

If the musical passage of the first half was from noise to speech to song, that of the second is from song to synthetic song to artificial timbre and back to noise with the explosion of Sophia. The first half is characterized by fragmentation, discontinuity and small-scale sound-events which accumulate to form dense ensemble textures. The second half is continuous music, with a vast linear sweep from the first entrance of the Lamptons to the coda.

The Lamptons enter first as the film fades out. They are singing a song about the search for 'Valis', using traditional electronic rock instruments and synthesizers as accompaniment. Their vocal style is classical but amplified:

> I want to see you, man.
> As quickly as I can.
> Let me hold your hand.
> I've got no hand to hold
> And I'm old, old; very old.
>
> Why don't you look at me?
> Afraid of what you see?
> I'll find you anyhow,
> Later or now; later or now.

The primary elements of the music which is made by the Lamptons are rhythmic definition and drive, and clear directional harmonic movement.

This accompanying music becomes more and more inharmonic and complex until the introduction of Mini, who prepares a performance with his enormous real-time system. This performance is the musical centrepiece of the opera. There are no words or text; all is generated by live timbral synthesis. Mini does not perform at a traditional keyboard, but rather manipulates an intelligent interactive real-time music system. A large library of musical materials is programmed into this system. Each of these materials can grow automatically when activated, into an ever-more-complicated web of music. Mini acts as a sort of super-conductor, shaping the overall complexity and structural direction of the music, while letting the live computer determine sonic details and individual events. This performance is a tour de force for Mini, and a real show-stopper. The contradictory human feelings of the beginning of the opera have been 'purified' and unified into a totally synthetic world (with no intrusion from human concerns): that of computer-generated sound.

But the moment of greatest unity comes in the next scene, the confrontation with Sophia. For here all the principles which I have been espousing in this chapter are pushed to their farthest extreme. All the diverse musical tendencies which have overlapped and conflicted during the whole course of the opera unite in Sophia. In fact, during the compositional process, I have conceived the music of Sophia first and then worked 'backwards' through the opera, deriving from Sophia's music the disjointed elements which will 'come together' here. For most of this scene, simple glorious singing is married to textual clarity, timbral delicacy, inharmonic accompaniment and synthetic instrument sounds. The main harmonic and melodic language of the piece is found in its most compact and refined form. As Sophia delivers her message ("You carry in you the voice and authority of wisdom; you are, therefore, Wisdom, even when you forget it. Be patient during this time; it will be a time of trials for you. But I will not fail you."), the attempt is to achieve the perfect balance between parameters, all tuned through computer synthesis, of which I spoke earlier. This point of pure equilibrium and pure 'music' is disrupted as Mini tampers with the Sophia 'hologram', injecting ever more complex synthetic timbre into the pure mixture, arriving at a point of saturation corresponding to Sophia's final annihilation!

The opera ends with the entire synthetic edifice vanishing into dust. The two human beings, Horselover Fat and Phil Dick are left to relate to each other and to the world with simple human speech, without the unifying aids of computer synthesis and transformation. We feel their loss, their isolation, but also respect their integrity and simplicity as they study the options still open to them in their search.

* * *

In order to achieve the musical objectives that I have set for myself in *Valis*, I have undertaken several research projects to develop computer music tools and concepts which do not yet exist.

I have made the point above about the necessity of unifying diverse materials into a coherent musical setting. In practical terms, this means using many different techniques for sound synthesis and transformation. At present, no unified system exists that allows a composer to control many different programs in a consistent way. Data must be specified differently for each program; often sounds cannot be sent from one part of a system to another since formats differ; physical or mathematical descriptions of musical structures must be given, rather than higher-level abstract concepts. This is why it is essential to conceive a coherent 'composer's work station'. Such an environment will include a homogeneous set of tools to permit the composer to navigate easily through the many processes necessary to make complex music. It will allow the composer to represent sound structures and events graphically. A large data base will permit the easy utilization and combination of pre-recorded sources, the manipulation of synthesis objects, and the digital signal processing of either. A unified set of descriptions will apply to synthesized *and* sampled sound treatment. Finally, an object-oriented, LISP-like structure editor will permit large musical forms to be described and controlled from a high level of abstraction. This work station is being designed for the individual composer and the fundamental operations will be possible on relatively inexpensive, commercially available technology. The composer will do all creative work, sketching, testing and so on in the quiet of his or her own studio. But the work station will be directly compatible with a large studio system, suitable for high quality final production or concert performance. The composer will bring pre-prepared data to this studio and perfect the final stages of a work with superior technology. All the vocal and general sound synthesis and transformation for *Valis* will be done with such a system.

To achieve the highest degree of real-time interaction between performers and a computer system, it is necessary to conceive a technology of 'Hybrid instruments'. Such a system will result in a complete union between performer and computer, rather than the relationship of adversary or partner which is common at present. A variety of microprocessors will work in parallel, connected directly to the live performer (in the case of my opera mostly singers). One set of processors will monitor physical characteristics of the performance, such as muscle tension, mouth position, air flow and breath pressure. A second set will perform real-time signal processing operations on the acoustic signal, such as pitch detection, spectral analysis, attack detection, phrase and density analysis and syntactic parsing of musical material. Active signal processing will also be performed, such as intelligent sonic transformations of the spectrum or musical phrases of the performer. A third set of processors will contain a set of generation programs, capable of creating original music. These processors will augment and amplify the live music performed, based on a musical understanding of this material. The key point is that this entire computer system will be automated, highly responsive, and directly controlled by the musical gestures and decisions of the live performer; no separate commands or instructions will be given to the machines. In a very real sense, the performer and computer will be *unified* into a new entity: the hybrid musical instrument. All of the live transformation of speech into noise, or speech into song from the opera will be achieved with such a technique, and would be virtually impossible to create otherwise.

The presence of such material will result in enormous differences between the formal development of *Valis* and that of *Spectres Parisiens*. In the latter the computer material was stretched out over time, adding structure and direction to the instrumental parts. In *Valis* the essence of the piece will be found in the perfectly balanced multi-parametric episode of the Sophia scene: each phrase, every note, will contain all of the material for the entire opera. Instead of growing forward in time, the elements fuse and implode towards this focal point of unification over the length of the entire opera.

A system of hybrid instruments will provide the context to create what I think of as 'intelligent musical agents'. The metaphor for this musical idea is drawn from Marvin Minsky's book *The Society of Mind*. As he states at the beginning of this book:

> To make a real theory of the mind, we have to show how minds are built from mindless stuff. This is because unless we can explain the mind in terms of things which have no thoughts or feelings of their own, we'll only have gone around in a circle. Our goal, therefore, must be to build minds from things which are much smaller and simpler than anything that we'd consider smart. I'll call these particles *agents*.[18]

This concept is of great importance in imagining an automated music system.

We are conceiving of programs in which multi-dimensional musical objects can be defined. Each of these objects will have information imbedded in it about all the musical parameters, about certain phrasing tendencies, and melodic and harmonic material. Each will be just a small fragment, but will contain a kind of genetic coding which will give it a tendency to grow as a cell, conscious of context and of other musical material that it might encounter while trying to expand. A protocol will be set up describing how these agents may interact and group together to create complex webs of musical material. Programs will define tendencies for growth, and a method be invented to allow the composer (or performer) to influence the path in which the music will unfold. Specific compositional or sonic details will be adjusted by the composer choosing to give attention to a particular level of the musical hierarchy. Plans are being made to implement such a system on one of the new generation of powerful parallel processing computers, such as the Connection Machine designed by Thinking Machines, Inc., in Cambridge, Massachusetts. Such computers are actually vast networks of individual processors – rather like intelligent registers – each of which can contain a multi-dimensional musical cell. The complex control structure of this machine enables the kind of intricate and ever-changing pattern of musical elements described here. This will constitute a living intelligent music system. It will be on such a system that Mini performs during his big 'number', and the resemblance of this concept to 'Valis' itself is only partly a coincidence.

A final aspect on which I am currently working, and which is central to the conception of the opera, is the aesthetic and technical unification of sound and image. Since the introduction of video technology and especially with the development of computer imaging, it has at last become possible to treat visual information in terms of forms and colours changing continuously over time – as electronic signals rather than discrete frames or stationary objects. This means that sound (which is by definition a time-based art form) and image can be conceived of in a similar framework. Although correspondences between the two arts should never be simplistic or obvious, the doors *have* been opened for a true synergistic marriage of the two. Neither need remain illustrative of the other. Similar formal patterns and structures can be expressed simultaneously in the two media. The integration of live sources with processed transformation and computer synthesis can be treated identically. This unification is aided by the fact that computer systems have been developed (by the Droid Works division of Lucasfilm,[19] for example) which, for the first time, permit the artist to create in a technologically homogeneous environment.

Music can be reinforced by the lucidity, clarity and directness that a sensitive visual support might provide. Visual art can be liberated and expanded by the transformational, ephemeral, and abstract model that music represents. In this marriage of music and time-variant image lies one of the greatest artistic adventures, and promises, of our age.

So, even in discussing computer art tools, I come back to the idea of unity: unity of conceptual resources, unity of technology and live performer, unity of musical result with structural description. And yet these ideas are still in the research stage, I am still looking, along with Horselover Fat and Phil Dick, for the unified world that I so greatly desire. In real life such a goal is surely many years off. In the more artificial world of music we are perhaps closer. Closer, at least, to being able to manipulate both computer and acoustic materials in a single musical context, with no difference between them, the one enriching the other, totally compatible, totally unified: one single artistic vision motivated by deep human concerns.

Notes

Chapter 2

1. As with Messiaen, who associates colours with several of his 'modes of limited transposition'. Various sources cited in Robert Sherlaw Johnson, *Messiaen* (London: Dent, 1975), p. 19, etc.
2. Trevor Wishart discusses this further in Chapter 3.
3. Since its foundation in 1948 by Pierre Schaeffer, this institution – both a studio and a group of composers – has had several names. It is referred to throughout this chapter as the Groupe de Recherches Musicales or GRM.
4. A phrase Ferrari used in conversation with Hansjörg Pauli, and quoted by the latter in his *Für wen komponieren Sie eigentlich?* (Frankfurt am Main: Fischer, 1971), p. 41.
5. Claude Lévi-Strauss, *The Raw and the Cooked* (London: Cape, 1970), p. 24.
6. These, and related discussions, are covered in such text-books as R. B. Braithwaite, *Scientific Explanation* (Cambridge: University Press, 1968), especially ch.9.
7. Pierre Schaeffer, *Traité des objets musicaux* (Paris: Seuil, 1966).
8. C. Hempel and P. Oppenheim, 'Studies in the Logic of Explanation' in *The Structure of Scientific Thought*, ed. E. H. Madden (Buffalo: University Press, 1960).
9. *Boulez on Music Today* (London: Faber and Faber, 1971), p. 31.
10. Ibid. p. 31.
11. E.g. *Plus–Minus*, *Prozession*, *Spiral*, *Kurzwellen*, *Pole*, *Expo*.
12. *Aus den sieben Tagen*, *Für kommende Zeiten*.
13. Boulez (see Note 9), p. 22.
14. *Study II* score, Universal Edition UE 12466, pp. IV–VI.
15. Seppo Heikinheimo, *The Electronic Music of Karlheinz Stockhausen* (Helsinki: Suomen Musiikkitieteelinen Seura, 1972), p. 152.

16. As given in Richard Kostelanetz, *John Cage* (New York: Praeger, 1970; and London: Allen Lane, 1971), pp. 109–11.
17. Two versions of *Fontana Mix* exist on disc: one 'fixed' on tape by the composer (available on Vox Turnabout), the other a recording of a 'live' version (*Fontana Mix/Feed*) by the percussionist Max Neuhaus (on Columbia).
18. IRCAM: the Institute de Recherche et Coordination Acoustique/Musique, part of the Pompidou Centre in Paris and directed since its establishment in 1974 by Pierre Boulez.
19. The term 'Stanford' is used to refer to the Center for Computer Research in Music and Acoustics, Stanford University, California.
20. MUSIC V is a music language for use on mainframe computers written by Max Mathews at Bell Telephone Laboratories in 1967–8. Originally used primarily for synthesis, subsequent versions have included extensive signal processing possibilities for recorded sounds.
21. CHANT is a program developed at IRCAM by Xavier Rodet, Yves Potard and Jean-Baptiste Barrière based on a model of the vocal tract and voice production. See their article 'The CHANT Project: From the Synthesis of the Singing Voice to Synthesis in General', *CMJ* [*Computer Music Journal*], VIII/3 (1984), pp. 15–31.
22. Further details are available in Jonathan Harvey, '*Mortuos Plango, Vivos Voco*: A Realization at IRCAM', *CMJ*, V/4 (1981), pp. 22–4.
23. Denis Smalley, 'Electroacoustic Music in Perspective', sleeve note accompanying the recording of *Pentes* (UEA Recordings UEA 81063).
24. Denis Smalley, 'Problems of Materials and Structure in Electro-acoustic Music', paper presented at the International Electronic Music Conference, Stockholm, 1981. Edited version reprinted in the *EMAS Newsletter*, IV/1 & 2 (London, 1982), pp. 3–8 & 4–7 respectively.
25. Originally published in the *Revue musicale*, no. 236 (Paris: Richard-Masse, 1957), but more recently in Pierre Schaeffer, *La musique concrète* (Paris: Presses Universitaires de France, 1967, 2nd ed. 1973), pp. 28–30.
26. The studio of RAI (Radio Audizioni Italiane, Studio di Fonologia), Milan was established between 1953 and 1955 and originally directed by Luciano Berio.
27. See for example Karl H. Wörner, *Stockhausen: Life and Work* (London: Faber and Faber, 1973), pp. 82–5.
28. As given, for example, in *Eine Schlüssel für Momente* (Kassel: Edition und Verlag Boczkowski, 1971), p. 9, and reprinted in Robin Maconie, *The Works of Karlheinz Stockhausen* (Oxford: University Press, 1976), p. 168.
29. Michael McNabb, '*Dreamsong*: The Composition', *CMJ*, V/4 (1981), p. 36.
30. Ibid. pp. 40–42.
31. Ibid. p. 36.
32. From the composer's programme note circulated with the GRM performance copy of the tape.

33. Pauli (see Note 4), p. 58.
34. Stockhausen, *Texte zur Musik 1963–1970*, III (Cologne: Dumont Schauberg, 1971), p. 272.
35. See the composer's comments in the introductory notes to *Hymnen* – on both the record sleeve (Deutsche Grammophon) and in the score (Universal Edition) – in which he explains that many aspects of the composition ". . . arose during my work on it, from the universal character of the material on which it is based . . .".
36. Jonathan Cott, *Stockhausen: Conversations with the Composer* (London: Robson, 1974), pp. 190–91.
37. Stockhausen (see Note 34), p. 80.
38. Trevor Wishart, *Red Bird: A Document* (published by the composer) is not a score in the accepted sense either for realization or for diffusion. It is an essay describing the philosophy, working method and material transformations of this 45-minute work.
39. Ibid. p. 6; but note that Wishart uses the term 'abstracted' in this quote to mean the same as the word 'abstract' in general use in this chapter.
40. Ibid. p. 7.
41. Pauli (see Note 4), p. 58.
42. The studios of Westdeutscher Rundfunk (WDR) in Cologne were founded in 1953. Originally the champions of the purely electronic synthesis of sound, though recorded material was integrated from about 1955 (beginning with the work cited). With the exception of *Telemusik*, all Stockhausen's electroacoustic works cited in this chapter were composed in this studio.
43. This common misconception (masquerading as a simplification) is found in most text-books of electroacoustic music which take an historical view.

Chapter 3

1. This chapter is based on Trevor Wishart's book *On Sonic Art* (York: Trevor Wishart, 83 Heslington Rd., 1985).
2. Pierre Schaeffer, *Traité des objets musicaux* (Paris: Seuil, 1966).
3. Luc Ferrari, in conversation with Hansjörg Pauli, and quoted in the latter's *Für wen komponieren Sie eigentlich?* (Frankfurt am Main: Fischer, 1971), p. 41.
4. CHANT was developed by Xavier Rodet, Yves Potard and Jean-Baptiste Barrière.
5. Trevor Wishart, *Red Bird* (York University Studio). Record from Trevor Wishart.
6. *On Sonic Art* (see Note 1).
7. Claude Lévi-Strauss, *The Raw and the Cooked* (London: Cape, 1970).
8. Murray Schafer, *The Tuning of the World* (New York: Knopf, 1977).

Chapter 4

1. This chapter is a comprehensive revision and extension of a paper originally presented at a conference on electroacoustic music organized by EMS (Elektronmusikstudion), Stockholm, in 1981.
2. The term 'spectro-morphology' is preferable to the Schaefferian term 'typomorphology' for the reasons given in the text. Pierre Schaeffer's *Traité des objets musicaux* (Paris: Seuil, 1966) is the first significant work to elaborate spectromorphological criteria, and it provides the foundations for this chapter.
3. See T. Georgiades, *Music and Language: The Rise of Western Music as Exemplified in Settings of the Mass*, trans. M. L. Göllner (Cambridge: University Press, 1982).
4. 'Reduced listening' (écoute réduite) is a Schaefferian concept. See Schaeffer (Note 2) and M. Chion, *Guide des objets sonores* (Paris: Buchet Chastel/INA GRM, 1983) for a full discussion.
5. The relationship between 'abstract' and 'concrete' is more complex than this discussion indicates. See Schaeffer (Note 2) and Chion (Note 4).
6. The 'cardinal' and 'ordinal' concepts are Schaeffer's. Further information can be found in Chion (see Note 4), pp. 43–8.
7. The qualities here attributed to 'gesture' and 'texture' can be found in a variety of writings, although not always clearly thought out. Stockhausen's associated concepts of 'gestalt', 'structure', 'stasis' and 'process' are close though muddled (see S. Heikinheimo, *The Electronic Music of Karlheinz Stockhausen*, Helsinki: Suomen Musiikkitieteelinen Seura, 1972, from p. 139). Boulez refers to 'gesture' and 'contemplation' in connection with the presentation of his work *Eclat* on the cassette *Le temps musical 1*, Radio France/IRCAM. Harvey discusses 'contemplation' in 'Reflection after Composition', *Contemporary Music Review*, ed. N. Osborne, I/1 (Harwood Academic Publishers, 1984), pp. 83–6.
8. No works have been referred to in this chapter. We suggest that no electroacoustic work is immune from the questions considered. Two works can be particularly recommended for study because of their comprehensive incorporation of the majority of spectro-morphological issues: Stockhausen's *Kontakte* and Bernard Parmegiani's *De Natura Sonorum*.

Chapter 5

1. Jan W. Morthenson in a paper delivered at the Third Colloquium of the Confederation Internationale de Musique Electroacoustique, Stockholm, 1986.
2. See D. Keane, 'Some Practical Aesthetic Problems of Electronic Music Composition', *Interface*, VIII (1979), pp. 196–7.
3. Noam Chomsky, *Syntactic Structures* (The Hague: Mouton and Co., 1957).
4. Benjamin Lee Whorf, *Language, Thought, and Reality* (Cambridge: MIT Press, 1956).

5. As exemplified by the following well-known titles: *Language of Art* (Nelson Goodman), *Art as the Language of the Emotions* (C. J. Ducasse), *The Language of Art and Art Criticism* (Joseph Margolis), *The Language of Music* (Deryck Cooke), *The Language of Modern Music.* (Donald Mitchell).
6. Jeremy Campbell, *Grammatical Man: Information, Entropy, Language, and Life* (New York: Simon & Schuster, 1982), p. 162 and (London, Allen Lane, 1983).
7. Jos Kunst, 'Making Sense in Music I: The Use of Mathematical Logic', *Interface*, V (1976), p. 5.
8. Bertrand Russell, *Selected Papers* (New York: Modern Library, 1927), p. 358.
9. From George H. Mead, *Mind, Self, and Society* (Chicago: University of Chicago Press, 1934).
10. From Gilbert King, 'What is Information?', *Automatic Control*, a collection of articles reprinted from *Scientific American* (New York: Simon & Schuster, 1955), pp. 83–6.
11. Neil Postman and Charles Weingartner, *Teaching as a Subversive Activity* (New York: Delacorte Press, 1966), p. 85.
12. Kunst (see Note 7).
13. Northrop Frye *et al.*, *The Harper Handbook to Literature* (New York: Harper & Row, 1985), p. 380.
14. Morthenson (see Note 1).
15. Leonard B. Meyer, *Emotion and Meaning in Music* (Chicago: University of Chicago Press, 1956), p. 271.
16. Quoted in Robert Scholes, *Structuralism in Literature: An Introduction* (New Haven: Yale University Press, 1974), pp. 83–4.
17. N. F. Dixon, *Preconscious Processing* (Chichester, England: John Wiley & Sons, 1981).
18. D. Keane, 'Architecture and Aesthetics: The Construction and Objectives of Elektronikus Mozaïk', *Proceedings of the International Computer Music Conference 1985* (San Francisco: Computer Music Association, 1985), pp. 199–206.
19. See Note 16.
20. Behavioural states in play/exploration after M. Hughes, 'Exploration and Play in Young Children', in *Exploration in Animals and Humans*, ed. J. Archer and L. Birke (Wokingham, England: Van Nostrand Reinhold, 1983), pp. 230–44.
21. See D. Keane, *Tape Music Composition* (London: Oxford University Press, 1980), pp. 103–17: 'Basic Aesthetic Considerations'.
22. O. E. Laske, 'Toward a Theory of Interfaces for Computer Music Systems', *CMJ*, I/4 (1977), pp. 53–60. See also O. E. Laske, 'Considering Human Memory in Designing User Interfaces for Computer Music', *CMJ*, II/4 (1978), pp. 39–45.

Chapter 6

1. I wish to direct the reader to Robert Erickson's excellent study of these matters in *Sound Structures in Music* (Berkeley: University of California Press, 1975).
2. Pierre Boulez, *Notes of an Apprenticeship*, trans. Herbert Weinstock (New York: Knopf, 1968), p. 231.
3. Moog, ARP, EMS, Buchla etc.
4. An interesting critical study of the use of synthesizers in the rock and jazz fusion appears in David Ernst's *The Evolution of Electronic Music* (New York: Schirmer, 1977), ch. 9: 'Synthesizers in Live Performance'.
5. 'Effects: Who Uses What?', *Studio Sound*, XX/9 (1979), pp. 58–74.
6. 'Effects, Reverbs and Equalizers', *Studio Sound*, XXVII/11 (1985), pp. 38–41.
7. Boulez (see Note 2), p. 220.
8. Such networks are referred to as Local Area Networks or LANs; for example: Appletalk, a mechanism for connecting several Apple Macintosh computers together so that each may share specialized resources like large disc systems or laser printers. See Gareth Loy, 'Musicians Make a Standard: The MIDI Phenomenon', *CMJ*, IX/4 (1986), pp. 8–26.
9. Roger Dannenberg and J. Bloch, 'Real-time Computer Accompaniment of Keyboard Performances', *Proceedings of the International Computer Music Conference 1985* (San Francisco: Computer Music Association, 1985), pp. 279–90.
10. See *The IMA Bulletin*, II/11 (1985), pp. 5–7; published by the International MIDI Association of North Hollywood, California, this issue contains a complete list of current MIDI software products designating each of the devices and computers for which they were written.

Chapter 7

1. *Webster's Seventh New Collegiate Dictionary* (Springfield, MA: G. & C. Merriam Co., 1981).
2. See Michael McNabb, '*Dreamsong*: The Composition', *CMJ*, V/4 (1981).
3. William Schottstaedt, 'Pla: A Composer's Idea of a Language', *CMJ*, VII/1 (1983). CCRMA: Center for Computer Research in Music and Acoustics, Stanford, California.
4. Benoit Mandelbrot, *Les objets fractals: forms, hasard, et dimension* (Paris: Flammarion, 1975); trans. into English as *Fractals, Forms, Chance, and Dimensions* (San Francisco: W. H. Freeman & Co., 1977).
5. Benoit Mandelbrot, *The Fractal Geometry of Nature* (San Francisco: W. H. Freeman & Co., 1982). The term fractal derives from the concept that these shapes have fractional dimension. For example, the type of curve made by a

coastline is somewhere between a line and a plane; if it were an infinitely squiggly line it would fill a plane. It could therefore have a fractal dimension of, say, 1.3.

6. R. F. Voss and J. Clarke, '1/f Noise in Music: Music from 1/f Noise', *Journal of the Acoustic Society of America*, LXIII (1978), pp. 258–63, and '"1/f Noise" in Music and Speech', *Nature*, CCLVII (London: Macmillan, 1975), pp. 317–8.

Chapter 8

1. B. Truax, *Acoustic Communication* (Norwood, NJ: Ablex, 1984).
2. B. Truax, 'A Communicational Approach to Computer Sound Programs', *JMT*, XX (1976), pp. 227–300.
3. B. Truax, 'The Inverse Relation between Generality and Strength in Computer Music Programs', *Interface*, VI (1980), pp. 1–8.
4. See Max V. Mathews *et al.*, *The Technology of Computer Music* (Cambridge, MA: MIT Press, 1969).
5. W. Buxton *et al.*, 'The Evolution of the SSSP Score-editing Tools', in C. Roads and J. Strawn, eds., *Foundations of Computer Music* (Cambridge, MA: MIT Press, 1985). Originally published in *CMJ*, III/4 (1979), pp. 14–25.
6. W. Buxton *et al.*, 'The Use of Hierarchy and Instance in a Data Structure for Computer Music', in Roads and Strawn (see Note 5). Originally published in *CMJ*, II/4 (1978), pp. 10–20.
7. B. Truax, 'The PODX System: Interactive Compositional Software for the DMX-1000', *CMJ*, IX/1 (1985), pp. 29–38.
8. See C. Roads, 'Grammars as Representation for Music', in Roads and Strawn (see Note 5; originally published in *CMJ*, III/1 (1979), pp. 48–55) and S. R. Holtzmann, 'Using Generative Grammars for Music Composition', *CMJ*, V/1 (1981), pp. 51–64.
9. See J. Myhill, 'Controlled Indeterminacy: A First Step Toward a Semistochastic Music Language', in Roads and Strawn (see Note 5; originally published in *CMJ*, III/3 (1979), pp. 12–14); L. Hiller, 'Composing with Computers: A Progress Report', *CMJ*, V/4 (1981), pp. 7–21; K. Jones, 'Compositional Applications of a Stochastic Process', *CMJ*, V/2 (1981), pp. 45–61; and T. Bolognesi, 'Automatic Composition: Experiments with Self-similar Music', *CMJ*, VII/1 (1983), pp. 25–36.
10. C. Ames, '*Crystals*: Recursive Structures in Automated Composition', *CMJ*, VI/3 (1982), pp. 46–64; 'Stylistic Automata in *Crystals*', *CMJ*, VII/4 (1983), pp. 45–56; and 'Applications of Linked Data Structures to Automated Composition', *Proceedings of the 1985 International Computer Music Conference* (San Francisco: Computer Music Association, 1985), pp. 251–8.
11. G. Englert, 'Automated Composition and Composed Automation', *CMJ*, V/4 (1981), pp. 30–35.
12. Described in O. Laske, 'Composition Theory in Koenig's Project One and Project Two', *CMJ*, V/4 (1981), pp. 54–65.

13. See X. Rodet and P. Cointe, 'FORMES: Composition and Scheduling of Processes', *CMJ*, VIII/3 (1984), pp. 32–50.
14. See D. Rosenboom and L. Polansky, 'HMSL (Hierarchical Music Specification Language): A Real-time Environment for Formal, Perceptual and Compositional Experimentation', *Proceedings of the 1985 International Computer Music Conference* (San Francisco: Computer Music Association, 1985), pp. 243–50.
15. J. Chadabe, 'Some Reflections on the Nature of the Landscape within which Computer Music Systems are Designed', CMJ, I/3 (1977), pp. 5–11.
16. J. Chadabe, 'Interactive Composing: An Overview', *CMJ*, VIII/1 (1984), pp. 22–7.
17. M. Bartlett, 'A Microcomputer-controlled Synthesis System for Live Performance', in Roads and Strawn (see Note 5; originally published in *CMJ*, III/1 (1979). pp. 25–9) and 'The Development of a Practical Live-performance Music Language', *Proceedings of the 1985 International Computer Music Conference* (San Francisco: Computer Music Association, 1985), pp. 297–302.
18. D. J. Collinge, 'Moxie: A Language for Computer Music Performance'. *Proceedings of the 1984 International Computer Music Conference* (San Francisco: Computer Music Association, 1984), pp. 217–20.
19. See B. Truax, 'The POD System of Interactive Composition Programs', *CMJ*, I/3 (1977), pp. 30–39.
20. See B. Truax, 'Timbral Construction in *Arras* as a Stochastic Process', *CMJ*, VI/3 (1982), pp. 72–7.
21. See B. Truax, 'The Compositional Organization of Timbre in a Binaural Space', *Proceedings of the 1983 International Computer Music Conference* (San Francisco: Computer Music Association, 1983).
22. J. C. Risset, 'The Musical Development of Digital Sound Techniques', in B. Truax and M. Battier, eds., *Computer Music* (Ottawa: Canadian Commission for Unesco, 1981), p. 129; see also J. C. Risset, 'Computer Music Experiments 1964–. . .', *CMJ*, IX/1 (1985), pp. 11–18.
23. The carrier waveforms are those designed to include harmonics close to the formant frequency of common vowels, a technique suggested by Chowning. See J. Chowning, 'Computer Synthesis of the Singing Voice', *Proceedings of the 1981 International Conference on Music and Technology* (Melbourne: La Trobe University, 1981).

Chapter 9

1. Boethius, *De institutione musica*, Book 1, section 2; trans. Oliver Strunk in *Source Readings in Music History* (New York: Norton, 1950), p. 85.
2. MUSIC V: see Max V. Mathews *et al.*, *The Technology of Computer Music* (Cambridge, MA: MIT Press, 1969) and J.-L. Richer, *Music V: manuel de reference* (Paris: IRCAM, 1979).

3. CHANT: see G. Bennett, 'Singing Synthesis in Electronic Music', *Research Aspects of Singing* (Stockholm: Royal Swedish Academy of Music, 1981), pp. 34–50; and X. Rodet *et al.*, 'The CHANT Project: From Synthesis of the Singing Voice to Synthesis in General', *CMJ*, VIII/3 (1984), pp. 15–31.
4. FORMES: see X. Rodet and P. Cointe, 'FORMES: Composition and Scheduling of Processes', *CMJ*, VIII/3 (1984), pp. 32–50.
5. Yamaha frequency modulation synthesis system, consisting in part of the QX1 sequencer, TX816 MIDI FM tone generation system, DX1 digital programmable algorithm synthesizer.
6. 4X real-time digital signal processor, developed at IRCAM by Guiseppe Di Giugno, licensed to Sogitec.
7. Curtis Roads, 'Interview with James Dashow' in *Composers and the Computer*, ed. Curtis Roads (Los Altos, CA: William Kaufmann, 1985), p. 36.
8. The following account is adapted from Jonathan Harvey, '*Mortuos Plango, Vivos Voco*: A Realization at IRCAM', *CMJ*, V/4 (1981), pp. 22–4.
9. See Jonathan Harvey *et al.*, 'Notes on the Realization of *Bhakti*', *CMJ*, VIII/3 (1984), pp. 74–8.
10. See Jan Vandenheede and Jonathan Harvey, 'Identity and Ambiguity: The Construction and Use of Timbral Transitions and Hybrids', *Proceedings of the 1985 International Computer Music Conference* (San Francisco, Computer Music Association, 1985), pp. 97–102.
11. In the case of the oboe and the Tibetan monks we were building on models already made at IRCAM by Xavier Rodet.
12. Jonathan Cott, *Stockhausen: Conversations with the Composer* (London: Robson, 1974), p. 75.
13. Vandenheede and Harvey (see Note 10), p. 6.
14. Igor Stravinsky, *Poetics of Music* (Cambridge: Harvard University Press, 1970), pp. 64–5.

Chapter 10

1. Philip Davis and Reuben Hersh, *The Mathematical Experience* (Boston: Houghton Mifflin, 1981), p. 172.
2. In A. P. French, ed., *Einstein: A Centenary Volume* (Cambridge: Harvard University Press, 1978), p. 235.
3. Ibid. p. 275.
4. Ibid. p. 65.
5. Geoffrey Chaucer, *Complete Works*, ed. W. Skeat (Oxford: University Press, 1973), 'Troilus and Criseyde', Book V, vv. 259–61.

6. Howard Gardner, *The Mind's New Science: A History of the Cognitive Revolution* (New York: Basic Books, 1985).
7. K. S. Robinson, *The Memory of Whiteness* (New York: TOR, 1985).
8. Ibid. p. 132.
9. Tod Machover, 'Un nouveau mode de composer', *Revue dialectique* (Paris: Dialectique-diffusion, 1979); 'Computer Music with and without Instruments', *Musical Thought at IRCAM* (London: Harwood Academic Press, 1984); *Quoi, quand, comment: la recherche musicale* (Paris: Christian Bourgois, 1985); 'Thoughts on Computer Music Composition' in C. Roads, ed., *Composers and the Computer* (Los Altos, CA: William Kaufmann, 1985); 'The Extended Orchestra', in J. Peyser, ed., *The Orchestra: Its Origins and Transformations* (New York: Charles Scribner's, 1986).
10. CRI Records SD506. Score published by Ricordi Editions, Paris.
11. To be released on Bridge Records in 1986. Score published by Ricordi Editions, Paris.
12. Philip K. Dick, *Valis* (London: Corgi Books, 1981).
13. Ibid. p. 215.
14. Ibid. pp. 15–16.
15. Ibid. p. 157.
16. From the opera libretto by Catherine Ikam, Tod Machover and Bill Raymond (in manuscript at time of publication).
17. Dick (see Note 12), pp. 104–5.
18. Marvin Minsky, *The Society of Mind* (New York: Simon & Schuster, 1985), p. 2.
19. Lucasfilm Ltd., San Rafael, California.

Selected Bibliography

This bibliography lists some of the more important references from the text and adds some items of relevance to the views put forward.

Appleton, J. and Perera, R. (eds.), *The Development and Practice of Electronic Music* (Englewood Cliffs, NJ: Prentice Hall, 1975)

Boulez, P., *Notes of an Apprenticeship*, trans. Herbert Weinstock (New York: Knopf, 1968)

——, *Boulez on Music Today* (London: Faber and Faber, 1971)

Chion, M., *Guide des objets sonores* (Paris: Buchet Chastel/INA/GRM, 1983)

Chion, M. and Reibel, G., *Les musiques electroacoustiques* (Aix-en-Provence: INA/ GRM Edisud, 1976)

Chomsky, N., *Syntactic Structures* (The Hague: Mouton and Co., 1957)

Cott, J., *Stockhausen: Conversations with the Composer* (London: Robson, 1974)

Erikson, R., *Sound Structure in Music* (Berkeley: University of California, 1975)

Harvey, J., *The Music of Stockhausen* (London: Faber and Faber, 1975)

Heikinheimo, S., *The Electronic Music of Karlheinz Stockhausen* (Helsinki: Suomen Musiikkitieteelinen Seura, 1972)

Keane, D., *Tape Music Composition* (Oxford: University Press, 1980)

Kostelanetz, R., *John Cage* (New York: Praeger, 1970 and London: Allen Lane, 1971)

Lévi-Strauss, C., *The Raw and the Cooked* (London: Cape, 1970)

McAdams, S., *Spectral Fusion and the Creation of Auditory Images* (Stanford, CA: University Press, 1980)

Manning, P., *Electronic and Computer Music* (Oxford: University Press, 1985)

Minsky, M., *The Society of Mind* (New York: Simon and Schuster, 1985)

Pauli, H., *Für wen komponieren Sie eigentlich?* (Frankfurt am Main: Fischer, 1971)

Pousseur, H., 'Calculation and Imagination in Electronic Music' in *Electronic Music Review* (1968), no. 5, p. 21.

Roads, C. (ed.), *Composers and the Computer* (Los Altos, CA: Kaufmann, 1985)

Ruwet, N., 'Contradictions within the Serial Language', *Die Reihe*, no. 6 (1964), pp. 65–76.

Schaeffer, P., *A la recherche d'une musique concrète* (Paris: Seuil, 1952)

——, *Traité des objets musicaux* (Paris: Seuil, 1966)

——, *La musique concrète* (Paris: Presses Universitaires de France, 1967, 2nd ed. 1973)

Schafer, R. M., *The Tuning of the World* (New York: Knopf, 1977)
Shepherd, J. *et al.*, *Whose Music? A Sociology of Musical Languages* (London: Latimer, 1977)
Stockhausen, K., 'The Concept of Unity in Electronic Music', in Boretz, B. and Cone, E. (eds.), *Perspectives on Contemporary Music Theory* (New York: Norton, 1972)
Truax, B., *Acoustic Communication* (Norwood, NJ: Ablex, 1984)
Wehinger, R., *Geörgy Ligeti: Artikulation* (Mainz: Schott, 1970) [listening score with disc and analysis]
Wishart, T., *Red Bird: A Document* (York: Wishart, 1978)
——, *On Sonic Art* (York: Imagineering Press, 1985)
Wörner, K., *Stockhausen: Life and Work* (London: Faber and Faber, 1973)

List of Works Cited

This list of works is not intended to be comprehensive; it lists works cited in the text whether electroacoustic or not. For electroacoustic items a recording is given wherever possible. If a score is also available the publisher is listed. For works which are not primarily electroacoustic recording details are omitted. For more complete listings readers are referred to Appleton and Perera (1975) and Manning (1985) and to M. Kondvacki *et al.*, eds., *International Electronic Music Discography* (Mainz: Schott, 1979)

Abbreviations: UE – Universal Edition; SV – Stockhausen Verlag; CCRMA – Center for Computer Research in Music and Acoustics, Stanford University; DG – Deutsche Grammophon; also *d* – disc, *cass* – cassette, *sc* – score, *st* – studio

Babbitt, M.,
Ensembles for Synthesizer (*d*: CBS MS 7051)
Boulez, P.,
Eclat (*sc*: UE 14283)
Structures (*sc*: UE 12267 and UE 13833)
Cage, J.,
Fontana Mix (*d*: Vox Turnabout TV34046S; *sc*: Peters P6712)
Williams Mix (*d*: Avakian 1 and Klett/Stuttgart 92422)
Carlos, W.,
Switched-on Bach (*d*: Columbia MS 7194)
Chafe, C.,
Solera (*st*: CCRMA)
Chowning, J.,
Stria (*cass*: CCRMA 1)
Ferrari, L.,
Hétérozygote (*d*: Philips 836885)
Music Promenade (*d*: Wergo 60046)

Presque Rien no. 1 (*d*: DG 2561 041)
Presque Rien no. 2 (*d*: INA/GRM 9104 fe)
Tautologos II (*d*: Disques BAM LD071)
Harvey, J.,
Bhakti (*sc*: Faber Music)
Inner Light (1) (*sc*: Novello)
Inner Light (2) (*sc*: Faber Music)
Inner Light (3) (*sc*: Novello)
Mortuos Plango, Vivos Voco (*d*: Erato STU 71544)
Song Offerings (*sc*: Faber Music)
Ligeti, G.,
Artikulation (*d*: Wergo 60059; *sc*: see Wehinger entry in bibliography)
Lucier, A.,
I Am Sitting in a Room (*d*: Source Record 3 in *Source Magazine*, No. 7, Sacramento, CA: 1970; *sc*: ibid. p. 70)
Lutosławski, W.,
Trois Poèmes d'Henri Michaux (*sc*: PWM)
Machover, T.,
Fusione Fugace (*sc*: Ricordi)
Light (*d*: CRI SD506; *sc*: Ricordi)
Nature's Breath (*sc*: Ricordi)
Spectres Parisiens (*d*: Bridge Records; *sc*: Ricordi)
String Quartet no. 1 (*sc*: Ricordi)
McNabb, M.,
Dreamsong (*d*: 1750 Arch Records S-1800 and Mobile Fidelity Sound Labs MFCD818)
Invisible Cities (*st*: CCRMA)
Nono, L.,
La Fabbrica Illuminata (*d*: Wergo 60038; *sc*: Ricordi)
Parmegiani, B.,
Dedans-Dehors (*d*: INA/GRM 9102 pa)
De Natura Sonorum (*d*: INA/GRM AM714.01)
Schaeffer, P.,
Etude aux Objets (*d*: Philips 6521 021 and INA/GRM 9106)
Schaeffer, P. and Henry, P.,
Symphonie pour un homme seul (*d*: Philips 6510 012)
Smalley, D.,
Pentes (*d:* UEA [University of East Anglia] Recordings UEA 81063)
Stockhausen, K.,
Aus den Sieben Tagen (*sc*: UE 14790)
Expo (*sc* :SV)
Für kommende Zeiten (*sc*: SV)
Gesang der Jünglinge (*d*: DG 138811)
Hymnen (*d*: DG 139 421/2; *sc*: UE 15142)
Kontakte (*d*: DG 138811, Wergo 60009 and Vox STGBY 638; *sc*: UE

13678 and UE 12426)
Kurzwellen (UE 14806)
Mikrophonie I (*d*: CBS 72647 and DG 2530 583; *sc*: UE 15138)
Mikrophonie II (*d*: CBS and DG 2530 583; *sc*: UE 15140)
Momente (*d*: DG 2709 055; *sc*: UE 13816)
Plus–Minus (*sc*: UE 13993)
Pole (*sc*: SV)
Prozession (*sc*: UE 14812)
Spiral (*sc*: UE 14957)
Study I (*d*: DG LP 16133)
Study II (*d*: DG LP 16133; *sc*: UE 12466)
Telemusik (*d*: DG 137 012; *sc*: UE 14807)

Subotnik, M.,
The Wild Bull (*d*: Nonesuch H71208)

Truax, B.,
Arras, on album *Androgyne* (*d*: Melbourne Recordings SMLP 4042/3)
Sequence of Earlier Heaven (*d*: Cambridge Street Records CSR 8501)

Wendt, L.,
From Frogs (*cass*: composer)

Wishart, T.,
Red Bird (*d*: York University Studio (UK) YES7; *sc*: see bibliography)